5 PSAT Math Practice Tests

Paul G Simpson IV

with the Staff of Test Professors

Library of Congress Control Number: 2011927888

ISBN: 978-0-9796786-6-0

ACKNOWLEDGMENTS

Thank you to all of our students, each of whom contributed to this book in his/her unique way. We wish we could list you all by name, but we would never have enough space. A special thanks to Terrence Park, who brought unparalleled thought and insight to this work. This book could not have been completed without the hard work and tirelessness of Summer Xia and David Hao.

5 PSAT MATH

PRACTICE TESTS

PSAT MATH PRACTICE TEST 1

Time-25 Minutes
20 Questions

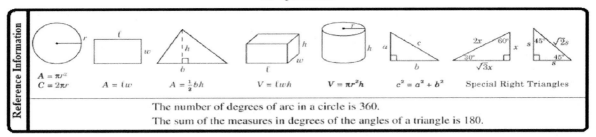

The number of degrees of arc in a circle is 360.
The sum of the measures in degrees of the angles of a triangle is 180.

1. What is x in terms of y if $2z = 3x$ and $3z = y$?

(A) $2/3y$
(B) $2/3z$
(C) $3z$
(D) $2y$
(E) $2/9y$

2. In a machine factory, one out of every 500 machines is defective. How many machines are <u>not</u> defective if the factory produced 20,000 machines?

(A) 40
(B) 400
(C) 19,060
(D) 19,960
(E) 19,996

3. The ratio of 1.75 to 1 is equal to which of the following ratios?

(A) 1 : 1.75
(B) 7 : 4
(C) 2 : 1.25
(D) 1 : 4
(E) 5 : 2

4. Points X (3, -6) and Y (5, 7) lie on the same line. Which one of the following may also lie on this line?

(A) (3, 45)
(B) (-2, -18)
(C) (-1, 32)
(D) (0, 51/2)
(E) (1, -19)

5. The average of x and y is also the mode of the set of x and y. If $y = 7$, what is the value of x?

(A) 5
(B) 6
(C) 7
(D) 8
(E) 9

6. If $|x - 5| = (3y)^2$ and $|y - 5| = 7$, where $x < 0$ and $y < 0$, what is the value of $y - x$?

(A) 12
(B) 29
(C) 33
(D) 36
(E) 41

7. The smallest integer of a set of consecutive integers is -8. If the sum of the integers is 19, what is the total number of integers in the set?

(A) 19
(B) 20
(C) 21
(D) 22
(E) 23

8. If line l passes through point (2, 5) and is perpendicular to the y-axis, what is the equation of line l?

(A) $y = 2$
(B) $y = x + 3$
(C) $y = x - 3$
(D) $y = 5$
(E) $y = x + 7$

9.

$$r, 1, s, t, \ldots$$

In the sequence above, r is the first term. Each term after the first term is five less than three times the preceding term. What is the value of rst?

(A) -836
(B) -44
(C) -11
(D) 15
(E) 44

10. If $x^{1/2} = 6$ and $\dfrac{x}{4} = \dfrac{3}{y}$, what is the value of y?

(A) 1/3
(B) 2/3
(C) 1
(D) 3
(E) 4

11. If the median of a set of five consecutive integers is equal to 14, then what is the value of the square root of the largest of these integers?

(A) $2\sqrt{3}$
(B) $\sqrt{13}$
(C) 4
(D) $\sqrt{15}$
(E) $\sqrt{14}$

12. A store determines its retail price by marking up the wholesale price by 60%. If an employee receives 20% off the retail price of an item and spends $80, then what is the wholesale price of the same item?

(A) $40.00
(B) $48.00
(C) $62.50
(D) $64.00
(E) $80.00

13. If six less than three times negative y is greater than or equal to three more than the square of nine, what is the value of y?

(A) $y \leq -30$
(B) $y \geq -30$
(C) $y \geq 27$
(D) $y \leq 33$
(E) $y \geq 36$

14. $2^a \times 2^b \times 2^c = 32$. If a, b, and c are positive integers where $b = a$, what is the highest possible value of ac?

(A) -2
(B) 0
(C) 1
(D) 2
(E) 3

15. John can take 4 tests per hour. Frank can take 10 tests per hour, and Ono can take 8 tests per hour. If Ono and Frank take tests together for one hour, and then John joins them, how many <u>minutes</u> will it take them to complete 73 tests?

(A) 150
(B) 180
(C) 200
(D) 210
(E) 230

16. A sack of marbles contains either blue or black marbles. The probability of choosing 2 blue marbles in a row without replacement is 3/10. If there is a total of 5 marbles in the sack, how many black marbles are there in all?

(A) 0
(B) 1
(C) 2
(D) 3
(E) 4

17. If the three sides of a triangle have integral values and two of the sides are 6 and 11. The third side can have how many distinct values?

(A) 1
(B) 11
(C) 12
(D) 17
(E) Infinitely many

18.

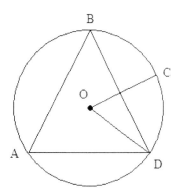

In the figure above, equilateral triangle ABD is inscribed in circle O. If arc BC = arc CD and OC = 6, what is the area of the sector of the circle bounded by OC and OD?

(A) 6π
(B) 12π
(C) 24π
(D) 30π
(E) 36π

19. If $f(x) = x^2 - 3$, then what is the value of $\dfrac{1}{3}(f(6) - f(3))$?

(A) 3
(B) 6
(C) 9
(D) 27
(E) 33

20.

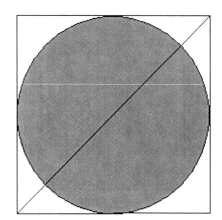

The shaded circle is inscribed in the square and the length of the diagonal is $\sqrt{72}$. What is the area of the shaded circle?

(A) 6π
(B) 9π
(C) 12π
(D) 36π
(E) $3\sqrt{2}\pi$

SECTION END

Time-25 Minutes
18 Questions

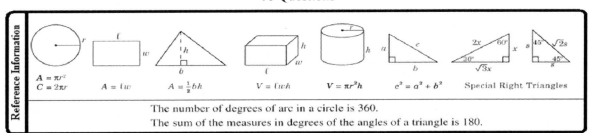

The number of degrees of arc in a circle is 360.
The sum of the measures in degrees of the angles of a triangle is 180.

1. A type of omelet includes only eggs, cheese, and mushrooms in the ratio of 20:3:9. If Diem wants to make a 64-ounce omelet, how many ounces of cheese does she need?

(A) 3
(B) 5
(C) 6
(D) 18
(E) 40

2. What is the value of t if $3t + 8 = 4t - 3$?

(A) 10
(B) 11
(C) 12
(D) 13
(E) 14

3. x and y are positive consecutive even integers whose sum is 70. If $x > y$, then what is the value of $x^{1/2}$?

(A) 6
(B) 34
(C) 35
(D) 36
(E) 1296

4. Joshua wants a computer that retails for $400. He can either take a 20% discount and pay $30 for shipping, or he can first receive 10% off and then use a coupon for an additional 10% off the computer with no shipping fee. How much cheaper is it to use the less expensive option?

(A) $20
(B) $24
(C) $26
(D) $324
(E) $350

5. If $(x + y)^2 = 49$, and $xy = 6$, what is the value of $x^2 + y^2$?

(A) -7
(B) 7
(C) 12
(D) 36
(E) 37

6. Five people are each asked to think of one negative integer. If the average of these integers is -21, what is the least possible value of one of the integers?

(A) -105
(B) -101
(C) -95
(D) -20
(E) -21

7. Jack base-jumped off a cliff that was 857 feet in altitude. The first second he fell two feet, and every second thereafter he fell three feet more than twice as many feet as the previous second. If he fell for 7 seconds before he released his parachute, at what altitude did he release his parachute?

(A) 243
(B) 254
(C) 603
(D) 614
(E) 830

8.

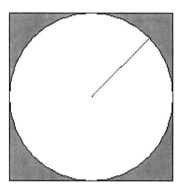

The radius of the inscribed circle is 3. What is the area of the shaded region?

(A) $9\pi - 18$
(B) $36 - 9\pi$
(C) $36 - 6\pi$
(D) $6\pi + 9$
(E) $24 + 6\pi$

9. A mixture contains 1/2 cup of flour, and equal parts of sugar, baking powder, and corn meal. If the mixture is 2 cups total, what is the ratio of the sugar to the total mixture?

10. If $x^3 = -64$ and $y^2 = 36$, what is the greatest possible value of $x + y$?

11. If the average of s and t is 80, and the average of s, t, u, and v is 56. What is the value of the square root of $(u + v)$?

12. If the graphs of $y = -3x + 4$ and $y = 2\sqrt{x} - z$ intersect at point $(4, c)$, what is the value of z?

13. Amy, Joe, Dan, Sally, and Lisa line up for a rollercoaster ride. If Joe is neither first nor last in line, and is never directly in front of or behind Amy, how many different arrangements are possible?

14. Hose C can fill a tanker truck in 6 hours and hose D can fill the same truck in 9 hours. How many hours will it take the hoses to fill the truck if they start at the same time and work together?

15. Clark brought a new fish tank that is 4 feet long, 2 feet wide and 1.5 feet high. He filled the tank with water 1 foot high. He added the 6 fishes he purchased to the tank and the water level rose to 1.2 feet. What is the approximate volume of the largest fish if the largest fish is twice as big as the next largest fish and so on? (Round to the nearest hundredth.)

16.

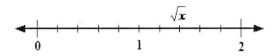

The marks on the number line above are equally spaced. What is the value of *x*?

18.

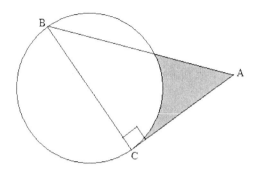

The length of $\overline{AB}$ is 15. If the length of $\overline{BC}$ is 3/5 the length of $\overline{AB}$, what is the area of $\triangle ABC$?

17. Stan buys *e* eggs from the store each week. If each egg costs *c* cents, what equation represents how many cents Stan spends on eggs per day in terms of *e* and *c*?

TEST END

Practice Test 1
Answer Key and Explanations

Section 1

1. E
2. D
3. B
4. E
5. C
6. B
7. A
8. D
9. E
10. A
11. C
12. C
13. A
14. E
15. D
16. C
17. B
18. A
19. C
20. B

Section 2

1. C
2. B
3. A
4. C
5. E
6. B
7. A
8. B
9. 1/4 or 1:4
10. 2
11. 8
12. 12
13. 36
14. 3.6
15. 0.81
16. 1.96
17. $(ec)/7$
18. 54

Finding Your Score

Raw score: Total Number Right – [Total Number Wrong ÷ 4] = _____

Notes: 1) For #9 - #18 in Section 2, only right answers are counted
2) Omissions are not counted towards your raw score
3) If the total number wrong ÷ 4 ends in .5 or .75, it is rounded up

Math Scoring Table			
Raw Score	Scaled Score	Raw Score	Scaled Score
38	80	15	46
37	76	14	45
36	72	13	44
35	70	12	43
34	69	11	42
33	68	10	40
32	66	9	39
31	65	8	38
30	64	7	36
29	63	6	35
28	61	5	34
27	60	4	33
26	59	3	31
25	57	2	29
24	56	1	28
23	55	0	26
22	54	-1	24
21	53	-2	22
20	51	-3	20
19	50	-4	20
18	49	-5	20
17	48	-6	20
16	47	-7	20

Strength and Weakness Review

Go back to the test and circle the questions that you answered incorrectly. This review will allow you to see what answer explanations to study more closely for problem-solving techniques. It will also allow you to see what question types you need to review more carefully.

	Section 1	Section 2
Integers / Consecutive Integers	7	6
Even / Odd Integers		3
Number Lines		16
Absolute Values	6	
Sequences	9	
Exponents		10
Fractional / Negative Exponents	10	
Word Problems	2, 13	7
Squaring and Square Roots	14	
Average, Median, and Mode	5, 11	11
Equations with 1 or 2 Variables		2, 5
Equations with 3 or 4 Variables	1	
Inequalities	13	
Building Equations		17
Slope		
X-Y Plane Problems	4, 8	12
Functions (Plug-Ins)	19	
Functions (Graphs & Shifts)		
Percents	12	4
Ratios	3	1, 9
Proportions		
Rates	15	14
Probability	16	
Combinations		13
Rectangles & Squares	20	
Angles		
Triangles	17	18
Special Triangles	18	
Other Polygons		
Circles	18, 20	8, 18
Volume		15

Section 1

1. (E) 2/9y
$2z = 3x \rightarrow z = 3/2x$
Since $3z = y$, $(3)(3/2x) = y \rightarrow 9x = 2y \rightarrow$
$x = 2/9y$

2. (D) 19,960
Defective = 1/500 machines
If there are 20,000 machines, then there are
$(1/500)(20,000) = 40$ defective machines.
Since the question asks for <u>not</u> defective
machines, the answer is $20,000 - 40 = \mathbf{19,960}$

3. (B) 7:4
To test, divide the ratios.
Since $7/4 = 1.75$, the ratio $7:4 = 1.75:1$

4. (E) (1, -19)
Points on the same line must have the same
slope, so first find the slope using X and Y:
$(7 - -6) / (5 - 3) \rightarrow 13/2$
Only $(1, -19)$ has the same slope, as proven by
using point X: $(-6 - -19) / (3 - 1) \rightarrow 13/2$

5. (C) 7
The mode is the most frequently occurring
number in a set. Since $y = 7$, x must also equal 7
in order for there to be a mode.

6. (B) 29
$y < 0$, so $|y - 5| = 7$ means that $y = -2$
$|x - 5| = (3y)^2 \rightarrow |x - 5| = (-6)^2 \rightarrow$
$|x - 5| = 36 \rightarrow x = -31$
$y - x = -2 - (-31) = \mathbf{29}$

7. (A) 19
The total number of integers from -8 to 8 is 17
(don't forget that 0 is an integer). The sum of
these integers = 0. Since the sum must equal 19,
we must add $9 + 10$. So the final number of
integers is 17 (from -8 to 8) + 2 (9 and 10) = **19**.

8. (D) y = 5
Because line l is perpendicular to the y-axis
where $y = 5$, it will always be a line for
which $y = \mathbf{5}$.

9. (E) 44
Sequence rule = 3(previous term) – 5
$r \rightarrow 3r - 5 = 1 \rightarrow 3r = 6 \rightarrow r = 2$
$s \rightarrow s = (3)(1) - 5 = 3 - 5 \rightarrow s = -2$
$t \rightarrow t = (3)(-2) - 5 = -6 - 5 \rightarrow t = -11$
$rst = (2)(-2)(-11) = \mathbf{44}$

10. (A) (1/3)
$x^{1/2} = 6 \rightarrow x = 36$, so $36/4 = 3/y \rightarrow 36y = 12 \rightarrow$
$y = \mathbf{1/3}$

11. (C) 4
Five consecutive integers: __, __, 14, __, 16
16 is largest integer $\rightarrow \sqrt{16} \rightarrow \mathbf{4}$

12. (C) $62.50
wholesale price = w
retail price + markup = 1.6
discount = 0.8
sales price = 80

$(w)(1.60)(.80) = 80 \rightarrow 1.28w = 80 \rightarrow w = \mathbf{62.5}$

13. (A) $y \le -30$
$(3)(-y) - 6 \ge 9^2 + 3 \rightarrow -3y - 6 \ge 84 \rightarrow -3y \ge 90$
$\rightarrow y \le \mathbf{-30}$

14. (E) 3
$2^a \times 2^b \times 2^c = 32$ and $2^5 = 32$, so $a + b + c = 5$
Since they are positive integers and $a = b$, the greatest possible value of ac occurs when $a = 1$, $b = 1$, and $c = 3$. In this case $ac = (3)(1) = \mathbf{3}$

15. (D) 210
1st Hour: Ono + Frank take $8 + 10$, or 18 tests
55 tests remain, so 55 / (22 tests / 60 min) =
$(55/22)(60) = 150$
1st hour + 150 minutes = **210 minutes**

16. (C) 2
Blue = x
First draw = $x/5$, and
Second draw = $(x - 1)/(5 - 1)$
So, $(x/5)((x - 1)/4) = 3/10 \rightarrow (x^2 - x)/20 = 6/20$
So, $x^2 - x = 6 \rightarrow x^2 - x - 6 = 0 \rightarrow (x - 3)(x + 2)$
Since x cannot be negative, $x = 3$
Black equals Total − Blue = $5 - 3 = \mathbf{2}$

17. (B) 11
Given two sides of a triangle, the third side is restricted to have, at least, the difference between the two sides and, at most, the sum of the two sides or: $11 - 6 < 3^{rd}$ side $< 11 + 6$. Therefore, $5 < 3^{rd}$ side < 17, and there are **11** distinct integer values between 5 and 17.

18. (A) 6π
Since OC = 6 and is a radius, the area of circle
$O = 36\pi$
Since triangle ABD is equilateral,
arc AB = arc BD = arc DA, so each arc covers 1/3 of the whole circle.
Since arc BD = 1/3 of the circle and arc BC = arc CD, the arc CD is equal to $(1/2)(1/3)$ of the circle, which equals 1/6.
Thus the area of the circle bounded by OC and OD equals 1/6 of the area of the circle, which equals $(1/6)(36\pi) = \mathbf{6\pi}$

19. (C) 9
$f(x) = x^2 - 3$, so
$f(6) = 6^2 - 3 = 33$ and $f(3) = 3^2 - 3 = 6$
$1/3(f(6) - f(3)) \rightarrow 1/3(33 - 6) \rightarrow 1/3(27) = \mathbf{9}$

20. (B) 9π
Diagonal = $\sqrt{72}$ or $(\sqrt{36})(\sqrt{2}) = 6\sqrt{2}$
Since the sides of the square form a 45-45-90 triangle, each side of the square is equal to 6.
Since side length of square = diameter of circle, the radius of the circle is equal to 3.
So the area of the circle = $\pi r^2 = \pi 3^2 = \mathbf{9\pi}$

Section 2

1. (C) 6
Cheese is 3 parts out of 32 total parts (3/32).
Since the total is doubled to 64, the parts of cheese must also be doubled: $(3)(2) = \mathbf{6}$

2. (B) 11
$3t + 8 = 4t - 3 \rightarrow -t = -11 \rightarrow t = \mathbf{11}$

3. (A) 6
Since $x > y$, $y = x - 2$
$x + x - 2 = 70 \rightarrow 2x = 72 \rightarrow x = 36 \rightarrow$
$x^{1/2} = \sqrt{x} = \sqrt{36} = \mathbf{6}$

4. (C) 26
Option 1: ($400)(.8) + $30 → 320 + 30 = $350
Option 2: ($400)(.9)(.9) → (360)(.9) = $324
So, option 2 is cheaper by **$26**.

5. (E) 37
$x + y = 7$ and $xy = 6$, so $x = 6$ and $y = 1$
$x^2 + y^2 = (6)^2 + (1)^2 = 36 + 1 = \mathbf{37}$

6. (B) -101
Since the average of the five integers is -21, the sum of the integers is -105. For four numbers to have the largest possible value, the other number must be as small as possible. Since the integers must be negative, the greatest possible value is -1. So, $-1 + -1 + -1 + -1 + x = -105 \rightarrow$
$x = \mathbf{-101}$

7. (A) 243
Feet traveled in 7 seconds =
$2 + 7 + 17 + 37 + 77 + 157 + 317 = 614$ feet
Since he started at 857 feet, he released his parachute at 857 feet – 614 feet = **243 feet**

8. (B) $36 - 9\pi$
If the radius of the inscribed circle equals 3, then its diameter equals 6, which is also the value of the sides of the square.
So the area of the square = (6)(6) = 36
The area of the circle $= \pi r^2 = \pi 3^2 = 9\pi$
Area of the shaded region = **36 – 9π**

9. 1/4 or 1:4
There are 4 total equal parts. So the ratio of the sugar (1 part) to the total mixture (4 parts) is 1:4.

10. 2
$x^3 = -64$, so greatest x-value is -4
$y^2 = 36$, so greatest y-value is 6
$x + y = -4 + 6 = \mathbf{2}$

11. 8
$(s + t)/2 = 80 \rightarrow s + t = 160$
$(s + t + u + v)/4 = 56 \rightarrow (160 + u + v) = 224 \rightarrow$
$u + v = 64$
So the square root of $u + v = \mathbf{8}$

12. 12
$-3x + 4 = 2\sqrt{x - z} \rightarrow -3(4) + 4 = 2\sqrt{4 - z} \rightarrow$
$-8 = 4 - z \rightarrow -z = -12 \rightarrow z = \mathbf{12}$

13. 36
There are five total spots in line.
Joe has three possible spots (in the middle).
Amy has only two possible spots because she can't be directly in front of or behind Joe.
Dan then has three possible spots.
Sally then has two possible spots.
Lisa then has only one possible spot.

Example: D̲ S̲ J̲ L̲ A̲

So the total possible combinations equal
(3)(2)(3)(2)(1) = **36**

14. 3.6

Hose C → 1 truck/ 6 hours = 1/6
Hose D → 1 truck / 9 hours = 1/9
Hose C + Hose D = 1/6 + 1/9 = 5/18
Total Job (1) → 1 / (5/18) = 18/5 = 3.6

15. 0.81

Initial Volume Water = lwh = (4)(2)(1) = 8
Volume Water with Fish = (4)(2)(1.2) = 9.6
So the volume of the fish = 9.6 – 8 = 1.6
There are 6 fish and x is the largest fish, so
$x + x/2 + x/4 + x/8 + x/16 + x/32 = 1.6$ →
$32x + 16x + 8x + 4x + 2x + x = 51.2$ →
$63x = 51.2$ → x = **0.81** (nearest hundredth)

16. 1.96

Since the marks are equally spaced, the value at
$\sqrt{x} = 1.4$
$\sqrt{x} = 1.4$ → $x = (1.4)^2$ → x = **1.96**

17. *ec*/7

cents / day = (~~eggs~~/week)(cents/~~egg~~) =
cents/week →
(cents/~~week~~) / (7 days/~~week~~) = *ec* / **7**

18. 54

If AB = 15, then BC = (3/5)(15) = 9 and
CA = 12
Area Triangle= $(1/2)bh$ = (1/2)(9)(12) = **54**

PSAT MATH PRACTICE TEST 2

Time-25 Minutes
20 Questions

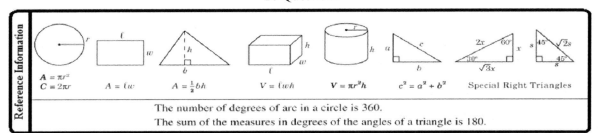

Reference Information

$A = \pi r^2$
$C = 2\pi r$
$A = \ell w$
$A = \frac{1}{2}bh$
$V = \ell wh$
$V = \pi r^2 h$
$c^2 = a^2 + b^2$
Special Right Triangles

The number of degrees of arc in a circle is 360.
The sum of the measures in degrees of the angles of a triangle is 180.

1. An auditorium has x rows that seat y students in each row. A senior assembly must seat z seniors. If 5 seats were left empty after the seniors sit, which of the following equations represents the relationship between x, y, and z?

(A) $xy = 5$
(B) $xy + z = 5$
(C) $xy - 5 = z$
(D) $xy + 5 = z$
(E) $xy = z - 5$

2. A certain store sells bottled water in cases of 18 and 24 bottles. If Joe bought three 18-bottle cases and one 24-bottle case, how many bottles of water did he purchase?

(A) 42
(B) 60
(C) 78
(D) 96
(E) 102

3. A certain box requires 18 people to lift. How many people are needed to lift three and a half boxes?

(A) 40
(B) 54
(C) 63
(D) 65
(E) 72

4. The least integer in a set of consecutive integers is -21. If the total sum of these integers is 45, how many total integers are in the set?

(A) 40
(B) 45
(C) 60
(D) 90
(E) 102

5. If $8 - p = -(r - 10)$ and $r = |-6|$, what is the value of p?

(A) 0
(B) 4
(C) 8
(D) 12
(E) 20

6.

297, 198,…

In the sequence above, each term after the first term, 297, is two-thirds of the preceding term. Which term is the first one that is not an integer?

(A) 5th
(B) 7th
(C) 12th
(D) 22nd
(E) 66th

7. If the average of w and v is 10, and the average of w and v and z is 9, what is the value of z?

(A) -11
(B) 1
(C) 7
(D) 17
(E) 19

8. c and d are negative consecutive even integers, and the absolute value of their sum is 34. If $c > d$, then what is the value of c^{-1}?

(A) -4
(B) -1/16
(C) 1/16
(D) 4
(E) 16

9.

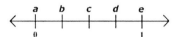

If there are four equal intervals between zero and one, which point represents the value of c^2?

(A) point a
(B) point b
(C) point c
(D) point d
(E) point e

10. Line l and line m (not shown) are perpendicular lines. If Line l contains the points (-2,-5) and (6, 9), what is the slope of line m?

(A) 7/4
(B) 4/7
(C) -1
(D) -4/7
(E) -7/4

11. If $t^{1/2} = (4u)^{1/4}$ and $u^{1/2} = 2^4$, then what is the value of t?

(A) 16
(B) 32
(C) 64
(D) 128
(E) 236

12. If $k = \dfrac{3y}{2t^2}$ and $y = 24$, then what is the value of t in terms of k?

(A) $\dfrac{36}{k}$

(B) $\dfrac{4\sqrt{3}}{k}$

(C) $\dfrac{6\sqrt{k}}{k}$

(D) $6\sqrt{k}$

(E) $36k$

13. If $3 \ge c \ge 0$ and $2 \ge d \ge -3$, what is the range of all the possible values of cd?

(A) $-9 \le cd \le 0$
(B) $-9 \le cd \le 6$
(C) $-6 \le cd \le 0$
(D) $-6 \le cd \le 6$
(E) $0 \le cd \le 6$

14.

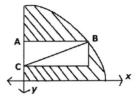

In the figure above, a rectangle is inscribed in a quarter-circle that lies in the xy-plane. If the arc of the quarter-circle is 6π, and the area of triangle ABC is 27, what is the area of the shaded area?

(A) $12\pi - 27$
(B) $12\pi - 36$
(C) $36\pi - 54$
(D) $36\pi - 27$
(E) $144\pi - 54$

15. A class has three spaces open and 8 applicants for the space. How many total combinations of new students are possible?

(A) 23
(B) 24
(C) 56
(D) 336
(E) 512

16. If *z* divided by negative thirty-two is greater than or equal to $(4)^{1/2}$, which of the following is the value of *z*?

(A) $z \le$ -64
(B) $z \le$ -16
(C) $z \ge$ -64
(D) $z \ge$ -64
(E) $z \le$ 64

17.

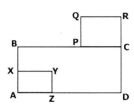

In the figure above, *X* is the midpoint of AB, *C* is the midpoint of RD, and *P* is the midpoint of BC. If the area of square AXYZ is 36 and the area of rectangle ABCD is 168, what is the perimeter of rectangle PQRC?

(A) 38
(B) 59
(C) 84
(D) 98
(E) 168

18. If $a \, ♪ \, b = (a-b)^2$, and $5 ♪ 3 = 7 ♪ a$, then which of the following is a possible value of *a*?

(A) -11
(B) -9
(C) 3
(D) 5
(E) 49

19. If $2^{y-2} = 16$ and $3^{z+2} = 243$, what is the value of $(yz)^2 - (y+z)^2$?

(A) 11
(B) 243
(C) 315
(D) 405
(E) 1600

20. If $h(k) = \dfrac{k^2}{2} + 3k + 4$ and $6h(k) = h(2k)$, then what is the product of all the possible values of *k*?

(A) 2
(B) 10
(C) 12
(D) 20
(E) 24

SECTION END

Time-25 Minutes
18 Questions

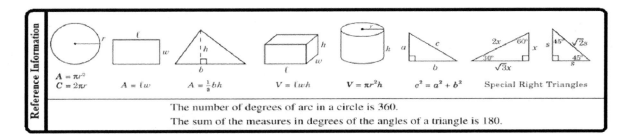

The number of degrees of arc in a circle is 360.
The sum of the measures in degrees of the angles of a triangle is 180.

1. A, B, and C are points on line segment AC, where A and C are endpoints. If BC is twice as long as AB, and AC = 24, what is the length of AB?

(A) 4
(B) 8
(C) 12
(D) 16
(E) 18

2. If the average of s and t is 12, and the average of s and t and u is 15. What is the value of u?

(A) -9
(B) 3
(C) 6
(D) 9
(E) 21

3. If $2^{x+1} = 16$ and $3^{y-3} = 27$, what is the value of $\dfrac{y}{x}$?

(A) 1/3
(B) 1/2
(C) 2
(D) 3
(E) 4

4.

$$x, 14, y, z, 133$$

In the sequence above, each term after the first term, x, is three more than twice the preceding term. What is the value of $x + z$?

(A) 31
(B) 63
(C) 70.5
(D) 86
(E) 96

5.

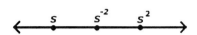

In the number line above, which of the following could represent the value of *s*?

(A) -2
(B) -1
(C) 0
(D) 1
(E) 2

---

6. If $\dfrac{(y + y + y)}{(y \times y \times y)} = 27$, and $y > 0$, then what is the value of *y*?

(A) -1/3
(B) -1/2
(C) 1/3
(D) 1/2
(E) 3

---

7. If $r = \dfrac{6s}{3t^2}$ and $s = 8$, then which of the following is the value of *t* in terms of *r*?

(A) $2/\sqrt{r}$
(B) $(2\sqrt{r}) / r$
(C) $(4\sqrt{r}) / r$
(D) $(14\sqrt{r}) / r$
(E) $(16\sqrt{r}) / r$

8.

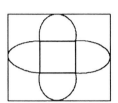

The central rectangle has a length of 4 and a width of 2, and it has 4 semicircles around its perimeter. If each of the semicircles is inscribed inside the outer square, what is the diagonal of the square?

(A) $2\sqrt{5}$
(B) 6
(C) $6\sqrt{2}$
(D) 36
(E) $36\sqrt{2}$

---

9. If 50 percent of 90 percent of *x* percent is 75 percent of 120 percent of *y*, what is the value of *x* in terms of *y*?

10.

$$a, 3a, \ldots$$

Each term after the first term, a, is three times the preceding term. If the sum of the first four terms is 80, what is the value of a?

11. Given that $|a - 2| = 9$ and $|w - 1| = 6$, where $a < 0$ and $w < 0$, what is the value of aw?

12. c and d are positive consecutive odd integers whose sum is 240. If $c > d$, then

$$\frac{1}{\sqrt{c}} = ?$$

13. If the sum of five consecutive positive integers is 80, what is the value of sum of the least of these integers and the median of these integers?

14. If $q^{1/2} = x$, $x^{-1/4} = 4^{-1/2}$, what is the value of q?

15. How many possible values exist for y if $4 - 3y < 10$, $-y - 8 > -14$, and y is an integer?

16. If e and f are inversely proportional and $e = 3$ when $f = 9$, then what is the value of e when $f = 3^{-1}$?

17. If $p^{1/3} = (8x)^{1/6}$, $x^{-1/2} = 4^{-1/4}$, and $4y = |\, p - 8|$, what is the value of y?

18. If $a(b) = b^2 - 2b$ and $4a(b) = a(2b) - 24$, what is the value of b?

TEST END

Practice Test 2
Answer Key and Explanations

Section 1

1. C
2. C
3. C
4. B
5. B
6. A
7. C
8. B
9. B
10. D
11. B
12. C
13. B
14. C
15. D
16. A
17. A
18. D
19. B
20. D

Section 2

1. B
2. E
3. C
4. C
5. A
6. C
7. C
8. C
9. $200y$
10. 2
11. 35
12. 1/11
13. 30
14. 256
15. 7
16. 81
17. 1
18. 6

Finding Your Score

Raw score: Total Number Right – [Total Number Wrong ÷ 4] = _____

Notes: 1) For #9 - #18 in Section 2, only right answers are counted
2) Omissions are not counted towards your raw score
3) If the total number wrong ÷ 4 ends in .5 or .75, it is rounded up

\multicolumn{4}{c}{**Math Scoring Table**}			
Raw Score	**Scaled Score**	**Raw Score**	**Scaled Score**
38	80	15	46
37	76	14	45
36	72	13	44
35	70	12	43
34	69	11	42
33	68	10	40
32	66	9	39
31	65	8	38
30	64	7	36
29	63	6	35
28	61	5	34
27	60	4	33
26	59	3	31
25	57	2	29
24	56	1	28
23	55	0	26
22	54	-1	24
21	53	-2	22
20	51	-3	20
19	50	-4	20
18	49	-5	20
17	48	-6	20
16	47	-7	20

Strength and Weakness Review

Go back to the test and circle the questions that you answered incorrectly. This review will allow you to see what answer explanations to study more closely for problem-solving techniques. It will also allow you to see what question types you need to review more carefully.

	Section 1	Section 2
Integers / Consecutive Integers	4, 8	12, 13
Even / Odd Integers		
Number Lines	9	5
Absolute Values		11
Sequences	6	4, 10
Exponents	19	3
Fractional / Negative Exponents	11	14, 17
Word Problems	2, 3	1, 6
Squaring and Square Roots		
Average, Median, and Mode	7	2, 13
Equations with 1 or 2 Variables	5	
Equations with 3 or 4 Variables	12	7
Inequalities	13, 16	15
Building Equations	1	
Slope	10	
X-Y Plane Problems		
Functions (Plug-Ins)	20	18
Functions (Graphs & Shifts)		
Percents		9
Ratios		
Proportions		16
Rates		
Probability		
Combinations	15	
Rectangles & Squares	14, 17	8
Angles		
Triangles		
Special Triangles		
Other Polygons		
Circles	14	8
Special Symbols	18	

Section 1

1. (C) $(xy) - 5 = z$
The total number of seats = xy
The total empty seats = 5
So z must equal total seats $- 5 \rightarrow z = (xy) - 5$

2. (C) 78
He bought three 18-bottle cases and one 24-bottle case, so the total bottles equals
$(3)(18) + (1)(24) = 54 + 24 = 78$

3. (C) 63
(18 people /1 box) : (x people / 3.5 boxes)
Cross-multiply: $x = (18)(3.5)$ and $x = 63$

4. (B) 45
The sum of the set of integers from -21 to 21 is zero. This set includes 43 integers, including 0. The next integers are 22 and 23, and the sum from -21 to 23 is equal to 45. So there are **45** total integers in the set.

5. (B) 4
$8 - p = -(6 - 10) \rightarrow 8 - p = -(-4) \rightarrow 8 - p = 4$
$\rightarrow p = 4$

6. (A) 5th term
Sequence rule = 2/3(previous term)
Terms: 297, 198, 132, 88, 58.67
The first term that is not an integer is the **5th term**, 58.67.

7. (C) 7
$(w + v) / 2 = 10 \rightarrow w + v = 20$
$((w + v) + z) / 3 = 9 \rightarrow 20 + z = 27 \rightarrow z = 7$

8. (B) -1/16
Since $c > d$, $d = c - 2$
$c + c - 2 = -34 \rightarrow 2c = -32 \rightarrow c = -16$
$c^{-1} = $ **-1/16**

9. (B) point b
Since there are four equal intervals, $c = 0.5$ and $c^2 = 0.25$
The value of 0.25 is represented by **point b**.

10. (D) -4/7
Slope of $l = (-5 - 9)/(-2 - 6) \rightarrow -14/-8 \rightarrow 7/4$
Since m and l are perpendicular, the slopes are negative reciprocals.
So, the slope of m is equal to **-4/7**

11. (B) 32
$u^{1/2} = 2^4 \rightarrow u^{(1/2)(2)} = 2^{(4)(2)} \rightarrow u = 2^8$
$t^{1/2} = (4u)^{1/4} \rightarrow t^{(1/2)(2)} = (4u)^{(1/4)(2)} \rightarrow t = (4u)^{1/2} \rightarrow$
$t = (2^2 \cdot 2^8)^{1/2} \rightarrow t = (2^{10})^{1/2} = 2^5 = $ **32**

12. (C) $(6\sqrt{k}) / k$
$$k = \frac{(3)(24)}{2t^2} \rightarrow k = \frac{72}{2t^2} \rightarrow k = \frac{36}{t^2} \rightarrow$$
$$t^2 = \frac{36}{k} \rightarrow t = \frac{\sqrt{36}}{\sqrt{k}} \rightarrow t = \frac{6}{\sqrt{k}} \rightarrow t = \frac{6\sqrt{k}}{k}$$

13. (B) $-9 \leq cd \leq 6$
$(3)(2) = 6$; $(-3)(3) = -9$; $(2)(0) = 0$; $(-3)(0) = 0$
Pick the lowest and highest values $\rightarrow$
$-9 \leq cd \leq 6$

14. (C) $36\pi - 54$

Circumference of quarter-circle = 6π, so the circumference of the entire circle is $(6\pi)(4) = 24\pi$.

$c = 2\pi r \rightarrow 24\pi = 2\pi r \rightarrow r = 12$; if $r = 12$, then the area of the whole circle is 144π and the area of the quarter circle is 36π.

Triangle ABC is half the area of the rectangle, so the area of the rectangle is $(17)(2) = 54$.
So the area of the shaded area is $36\pi - 54$.

15. (D) 336

1st position: 8 possibilities; 2nd: 7 possibilities; 3rd: 6 possibilities, so $(8)(7)(6) = \textbf{336}$

16. (A) $z \le -64$

$z / -32 \ge 4^{1/2} \rightarrow z / -32 \ge 2 \rightarrow z \le \textbf{-64}$

17. (A) 38

Since Area of AXYZ = 36, BX and XA = 6 and BA = 12

Since BA = 12 and Area of ABCD = 168, BC = 14

Since P is the midpoint of BC, PC = 7

Since PQRC is a rectangle, PC and QR = 7

C is the midpoint of RD, so CD and CR = 12

Since PQRC is a rectangle, CR and QP = 12

Total perimeter = PC + QR + CR + QP = **38**

18. (D) 5

$a \, \flat \, b = (a - b)^2$

$5 \, \flat \, 3 = (5 - 3)^2 = 2^2 = 4$

$7 \, \flat \, a = (7 - a)^2 = a^2 - 14a + 45$

$4 = a^2 - 14a + 45 \rightarrow a^2 - 14a + 41 = 0 \rightarrow$

$(a - 5)(a - 9) \rightarrow$ So $a = \textbf{5}$

19. (B) 243

$2^{y-2} = 16 \rightarrow 2^{y-2} = 2^4 \rightarrow y - 2 = 4 \rightarrow y = 6$

$3^{z+2} = 243 \rightarrow 3^{z+2} = 3^5 \rightarrow z + 2 = 5 \rightarrow z = 3$

$(yz)^2 - (y + z)^2 = [(6)(3)]^2 - (6 + 3)^2 = 324 - 81 = \textbf{243}$

20. (D) 20

$6h(k) = 6[(k^2/2) + 3k + 4] \rightarrow (6k^2/2) + 18k + 24$

$\rightarrow 3k^2 + 18k + 24$

$h(2k) = [(2k)^2/2] + (3)(2k) + 4 \rightarrow$

$(4k^2/2) + 6k + 4 \rightarrow 2k^2 + 6k + 4$

$6h(k) = h(2k) \rightarrow 3k^2 + 18k + 24 = 2k^2 + 6k + 4$

$\rightarrow k^2 + 12k + 20 \rightarrow (k + 10)(k + 2) \rightarrow$

$k = -2, -10 \rightarrow (-2)(-10) = \textbf{20}$

Section 2

1. (B) 16

BC = 2AB, so 2AB + AB = 24 $\rightarrow$ 3 AB = 24 $\rightarrow$ AB = **16**

2. (E) 21

$s + t = 24$, so $(s + t + u) / 3 = 15 \rightarrow$

$(24 + u) / 3 = 15 \rightarrow 24 + u = 45 \rightarrow \textbf{u = 21}$

3. (C) 2

$x \rightarrow x + 1 = 4 \rightarrow x = 3$

$y \rightarrow y - 3 = 3 \rightarrow y = 6$

$y/x = 6/3 = \textbf{2}$

4. (C) 70.5

x: $14 = 2x + 3 \rightarrow 2x = 11 \rightarrow x = 5.5$

z: $133 = 2z + 3 \rightarrow 2z = 130 \rightarrow z = 65$

$x + z = 5.5 + 65 = \textbf{70.5}$

5. (A) -2
Because $s < 1/s^2$, s must be a negative number.
This fact eliminates choices (C), (D), and (E).
Since $1/s^2$ must be less than s^2 plugging in
choice (B) produces a false result: $1 < 1$.
Thus all the given rules are only satisfied by **A
(-2)**.

6. (C) 1/3
$3y / y^3 = 27 \rightarrow 3/y2 = 27 \rightarrow 3 = 27y^2 \rightarrow$
$y^2 = 1/9 \rightarrow y = \sqrt{1/9} \rightarrow y = \mathbf{1/3}$

7. (C) 4 / $\sqrt{r}$
$r = (6)(8) / 3t^2 \rightarrow r = 48 / 3t^2 \rightarrow t^2 r = 16 \rightarrow$
$t^2 = 16 / r \rightarrow t = \sqrt{16} / \sqrt{r} \rightarrow t = \mathbf{4 / \sqrt{r}}$

8. (C) 6√2
Sides of exterior square = $2 + 2 + 2 = 6$
The diagonal is thus the hypotenuse of a 45-45-
90 triangle, so the diagonal = **6√2**

9. 200y
$(.5)(.9)(x/100) = (.75)(1.2)(y) \rightarrow .45x/100 = .9y$
$\rightarrow .45x = 90y \rightarrow x = \mathbf{200y}$

10. 2
First four terms: $a, 3a, 9a, 27a$
So, $40a = 80 \rightarrow a = \mathbf{2}$

11. 35
$a < 0$, so $| a - 2 | = 9$ means $a = -7$
$w < 0$, so $| w - 1 | = 6$ means $w = -5$
$aw = (-7)(-5) = \mathbf{35}$

12. 1/11
Since $d < c$, $d = c - 2$
$c + (c - 2) = 240 \rightarrow 2c - 2 = 240 \rightarrow 2c = 242$
$\rightarrow c = 121$
$1 / \sqrt{c} = 1 / \sqrt{121} = \mathbf{1 / 11}$

13. 30
Five consecutive integers: __, __, __, __, __
Average = $80 / 5 = 16$:   14, __, 16, __, __;
median = 16
(least integer) + (median) = (14) + (16) = **30**

14. 256
$x^{-1/4} = 4^{-1/2} \rightarrow x^{(-1/4)(-4)} = 4^{(-1/2)(-4)} \rightarrow x = 4^2 \rightarrow$
$x = 16$
$q^{1/2} = x \rightarrow q^{1/2} = 16 \rightarrow q = 16^2 \rightarrow q = \mathbf{256}$

15. 7
$4 - 3y < 10 \rightarrow -3y < 6 \rightarrow y > -2$
$-y - 8 > -14 \rightarrow -y > -6 \rightarrow y < 6$
So possible values are -1, 0, 1, 2, 3, 4, 5, for a
total of **7 possibilities**

16. 81
f decreases by a factor of 27, from 9 to 1/3
so, e <u>increases</u> by a factor of 27, from 3 to **81**

17. 1
$x^{-1/2} = 4^{-1/4} \rightarrow x^{(-1/2)(-2)} = 4^{(-1/4)(-2)} \rightarrow x = 4^{1/2} = 2$
$p^{1/3} = (8x)^{1/6} \rightarrow p^{(1/3)} = (16)^{1/6} \rightarrow p^{1/3} = (16)^{1/6} \rightarrow$
$p = 16^{1/2} = 4$
$4y = | p - 8 | \rightarrow 4y = | 4 - 8 | \rightarrow 4y = 4 \rightarrow y = \mathbf{1}$

18. 6

$a(b)_1 = b^2 - 2b$, and $4a(b) = a(2b) - 24$

$4a(b) = 4b^2 - 4b - 24 \rightarrow a(b)_2 = b^2 - b - 6$

$a(b)_1 = a(b)_2$, so $b^2 - 2b = b^2 - b - 6 \rightarrow -b = -6$

$\rightarrow b = \mathbf{6}$

PSAT MATH
PRACTICE TEST 3

Time-25 Minutes
20 Questions

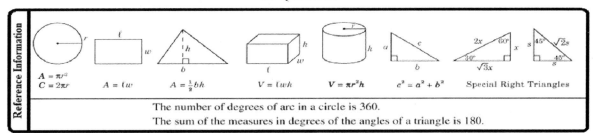

The number of degrees of arc in a circle is 360.
The sum of the measures in degrees of the angles of a triangle is 180.

1. 17 more than 7 times a number is equal to 101. What is the value of the number?

(A) 12
(B) 24
(C) 36
(D) 84
(E) 118

2.

$$256, 64, \ldots$$

In the sequence above, each term after the first term, 256, is one-fourth of the preceding term. Which term is the first one that is <u>not</u> an integer?

(A) 3rd term
(B) 4th term
(C) 5th term
(D) 6th term
(E) 7th term

3. A type of pasta includes starch, water, and salt in the ratio of 6:7:3. If the pasta weighs 72 total ounces, how many ounces of starch are contained in the pasta?

(A) 12
(B) 14
(C) 24
(D) 27
(E) 32

4. What is the value of $(-2x)^3 (6 - 3) + 3x^3$?

(A) $-21x^3$
(B) $-3x^3$
(C) 0
(D) $3x^3$
(E) $5x^3$

5. Hazel spends $72 on chocolate bars, which represents 16 percent of her total budget. If she spends 28 percent on SAT Prep books, how much does she spend on the books?

(A) $20
(B) $56
(C) $72
(D) $112
(E) $126

6. A line with a slope of -2/3 is reflected about the *y*-axis. What is the slope of the new line?

(A) -3/2
(B) -2/3
(C) 0
(D) 2/3
(E) 3/2

7. If $x > y > z > 0$, which of the following statements must be true?

I. $\dfrac{x-z}{x-y} > \dfrac{x-y}{x-z}$

II. $xy < xz$

III. $\dfrac{y}{z} > \dfrac{y}{x}$

(A) I only
(B) II only
(C) I and II
(D) I and III
(E) I, II, and III

8. If the ratio of boys to girls in a class is 11: 9, which one of the following could be the total number of girls in the class?

(A) 20
(B) 33
(C) 96
(D) 198
(E) 201

9.

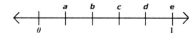

If there are five equal intervals between zero and one, which point is one-fifth the value of point *b* divided by point *a*?

(A) point *a*
(B) point *b*
(C) point *c*
(D) point *d*
(E) point *e*

10. If $-2 \times -|k|^3 = 3(l - 4)$ and $k = -3$, then what is the value of *kl*?

(A) -66
(B) -33
(C) 18
(D) 33
(E) 66

11. If *m* and *n* are positive consecutive integers whose sum is 17 and $1/m < 1/n$, what is the value of *m*?

(A) 6
(B) 7
(C) 8
(D) 9
(E) 10

12. If $(de + m)^2 = 9e^2 + ne + 49$, what is the value of n?

(A) 42
(B) 63
(C) 126
(D) 441
(E) 882

13. How many possible values exist for a if $14 - 3a < 23$, $a - 5 < -2$, and a is an integer?

(A) 2
(B) 3
(C) 4
(D) 5
(E) 6

14. If $a^{-2/3} = b^2$ and $b^{1/2} = c^{2/3}$, then what is the value of a in terms of c?

(A) c^{-4}
(B) c^{-2}
(C) $c^{1/2}$
(D) c^2
(E) c^4

15. A square and a circle have the same area. If the diagonal of the square is $6\sqrt{2}$, then what is the radius of the circle?

(A) $(3\sqrt{\pi})/\pi$
(B) $(6\sqrt{\pi})/\pi$
(C) $(9\sqrt{\pi})/\pi$
(D) $(27\sqrt{\pi})/\pi$
(E) $(36\sqrt{\pi})/\pi$

16. If $\dfrac{a^2 - a - 20}{a - 5} = b - 6 + a$, what is the value of b?

(A) -4
(B) -2
(C) 4
(D) 6
(E) 10

17. If the graphs of $y = -5x + 8$ and $y = 4\sqrt{x} - u$ intersect at point $(4, a)$, what is the value of u?

(A) -28
(B) -20
(C) 20
(D) 28
(E) 100

18. 40 percent of 60 percent of x is 40 percent of 10 percent of y. What is the product of the ratio of y to x and the ratio of x to y?

(A) 1
(B) 2
(C) 4
(D) 8
(E) 16

19.

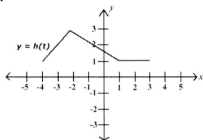

The above graph represents $y = h(t)$. Which of the following graphs best represents the graph of $y = h(t - 1) - 2$?

(A)

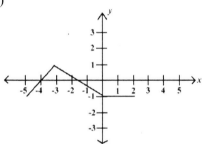

(B)

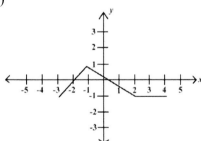

(C)

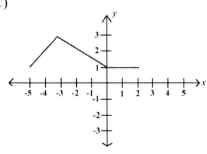

(D)

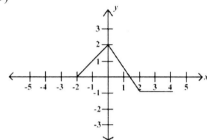

(E)

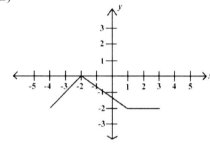

20. If $f(x) = \dfrac{x^2}{6} - 4$ and $2a(b) = (f(12) + 5)^{1/2}$, what is the value of $a(b)$?

(A) 1
(B) 1.5
(C) 2
(D) 2.5
(E) 5

SECTION END

Time-25 Minutes
18 Questions

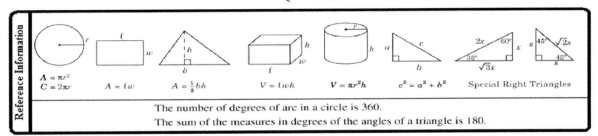

$A = \pi r^2$
$C = 2\pi r$ $A = lw$ $A = \frac{1}{2}bh$ $V = lwh$ $V = \pi r^2 h$ $c^2 = a^2 + b^2$ Special Right Triangles

The number of degrees of arc in a circle is 360.
The sum of the measures in degrees of the angles of a triangle is 180.

1. Ashley receives an allowance of $15 weekly. She receives an additional $2 for any homework assignment she completes. How much money does Ashley receive in one week if she completes 8 homework assignments?

(A) $16
(B) $23
(C) $25
(D) $31
(E) $120

2. If q is a positive integer and $q^2 = q^3$, then what is the value of q?

(A) -1
(B) -1/2
(C) 0
(D) 1/2
(E) 1

3. Ricardo's test results are 83, 92, 79, and 96. What is the lowest score he can achieve on the fifth and final test and still maintain an average of 90?

(A) 84
(B) 88
(C) 90
(D) 92
(E) 100

4. If m is directly proportional to n and $m = 3$ when $n = 6$, what is the value of n when $m = 12$?

(A) -3
(B) 3/2
(C) 15
(D) 18
(E) 24

5. If line p passes through point (-2, 3) and is perpendicular to the y-axis, what is the equation of line p?

(A) $y = -2$
(B) $y = -3/2$
(C) $y = 3/2$
(D) $y = 3$
(E) $y = 5$

6. If s and p^2 are inversely proportional and $s = 9$ when $p^2 = 6$, what is the value of p when $s = 6$?

(A) 2
(B) 3
(C) 4
(D) 6
(E) 9

7. If $\dfrac{k+k+k+k}{k \times k \times k \times k} = 256$ and $k > 0$, what is the value of k?

(A) 1/4
(B) 1/2
(C) 1
(D) 2
(E) 4

8.

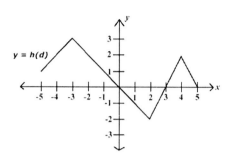

The above graph represents $y = h(d)$. If $h(d) = 3$ and $f(x) = 2h(d) - d$, what is the value of $f(x)$?

(A) 3
(B) 5
(C) 6
(D) 8
(E) 9

9. The height in feet, h, of a balloon after s seconds is determined by the function $h(s) = 60s - 3s^2$. What is the height of the balloon after four seconds?

10. Three people are each asked to think of one non-negative integer. If the average of these integers is 5, what is the least possible value of one of the chosen integers?

11. If $\sqrt{16x} = 8y^{-1}$ and x and y are positive integers, what is the value of x in terms of y?

12.

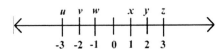

Which point best approximates the value of point x when it is squared and then divided by point w?

13. If $a = b^2c^2$, a is not equal to zero, and the values of b and c are doubled, then how many times is the value of a increased?

14. If Martina had 20 ferrets and now has 30 ferrets, what is the percent change in the number of her ferrets?

15. Two right triangles are joined together to form square WXYZ. If the hypotenuse of each of the triangles is $7\sqrt{2}$, then what is the area of square WXYZ?

16. If $b > b^2 > b^3$, which of the following statement(s) must be true?

I. $\dfrac{b^2}{b} > \dfrac{b}{b^3}$

II. $b^3 - b^2 < b - (b^2 \times b^3)$

III. $b^3 + \dfrac{b}{b^2} < b^{\frac{3}{2}}$

17. Winston bakes 4 pies per hour, and Clyde bakes 6 pies per hour. If Winston bakes alone for one hour, and then Clyde joins him, how many <u>minutes</u> will it take them to bake 32 pies?

18.

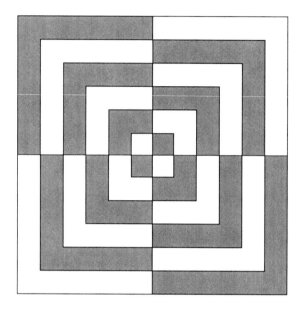

The square above has a side length of 6 and is divided into four smaller equal squares. What is the area of the shaded region?

TEST END

Practice Test 3
Answer Key and Explanations

Section 1

1. A
2. D
3. D
4. A
5. E
6. D
7. D
8. D
9. B
10. A
11. D
12. A
13. D
14. A
15. B
16. E
17. C
18. A
19. B
20. D

Section 2

1. D
2. E
3. E
4. E
5. D
6. B
7. A
8. E
9. 192
10. 0
11. $4/y^2$
12. w
13. 16
14. 50%
15. 49
16. II only
17. 228
18. 18

Finding Your Score

Raw score: Total Number Right – [Total Number Wrong ÷ 4] = _____

Notes: 1) For #9 - #18 in Section 2, only right answers are counted
2) Omissions are not counted towards your raw score
3) If the total number wrong ÷ 4 ends in .5 or .75, it is rounded up

Math Scoring Table			
Raw Score	**Scaled Score**	**Raw Score**	**Scaled Score**
38	80	15	46
37	76	14	45
36	72	13	44
35	70	12	43
34	69	11	42
33	68	10	40
32	66	9	39
31	65	8	38
30	64	7	36
29	63	6	35
28	61	5	34
27	60	4	33
26	59	3	31
25	57	2	29
24	56	1	28
23	55	0	26
22	54	-1	24
21	53	-2	22
20	51	-3	20
19	50	-4	20
18	49	-5	20
17	48	-6	20
16	47	-7	20

Strength and Weakness Review

Go back to the test and circle the questions that you answered incorrectly. This review will allow you to see what answer explanations to study more closely for problem-solving techniques. It will also allow you to see what question types you need to review more carefully.

	Section 1	Section 2
Integers / Consecutive Integers	11	
Even / Odd Integers		
Number Lines	9	12
Absolute Values		
Sequences	2	
Exponents	4	2, 13
Fractional / Negative Exponents	14	11
Word Problems	1, 5	1, 10, 13
Squaring and Square Roots		
Average, Median, and Mode		3
Equations with 1 or 2 Variables	10, 16	7, 11
Equations with 3 or 4 Variables	12	
Inequalities	7, 11, 13	6, 16
Building Equations		
Slope		5
X-Y Plane Problems	6	5
Functions (Plug-Ins)	20	9
Functions (Graphs & Shifts)	19	8
Percents	18	14
Ratios	3, 8	
Proportions		4, 6
Rates		17
Probability		
Combinations		
Rectangles & Squares	15	15, 18
Angles		
Triangles		
Special Triangles		15
Other Polygons		
Circles	15	
Special Symbols		

Section 1

1. (A) 12
$17 + 7x = 101 \rightarrow 7x = 84 \rightarrow x = 12$

2. (D) 6th Term
sequence: 256, 64, 16, 4, 1, **1/4**

3. (D) 27
The total parts of the ratio are $6 + 7 + 3 = 16$, so
6 (starch) / 16 (total) = x / 72 (total) $\rightarrow$
$16x = 432 \rightarrow x = 27$

4. (A) $-21x^3$
$(-2x)^3 (6-3) + 3x^3 \rightarrow -8x^3 (3) + 3x^3 \rightarrow$
$-24x^3 + 3x^3 = -21x^3$

5. (E) 126
B = total budget $\rightarrow .16b = 72 \rightarrow b = 450$
$.28b = (.28)(450) = 126$

6. (D) 2/3
When a line is reflected, the slope of the new
line is the negative of the original line. So
$(-2/3)(-) \rightarrow 2/3$

7. (D) I and III
Assume $x = 6, y = 4, z = 2$
I. $(6 - 2) / (6 - 4) > (6 - 4) / (6 - 2) \rightarrow 2 > 1/2$
III. $4/2 > 4/6 \rightarrow 2 > 2/3$

8. (D)
The total number of girls must be evenly
divisible by 9, since it is impossible to have a
fractional number of girls. So, the only number
that fits is **198**

9. (B) point b
$a = 1/5 = 0.2$ and $b = 2/5$ or 0.4
$x = (1/5)(b/a) = (1/5)(0.4/0.2) = 0.4$, which is
equal to **point b**

10. (A) -66
Since $k = -3$, $(-2) (|-3|^3) = 3(l - 4) \rightarrow$
$(-2)(-27) = 3l - 12 \rightarrow 54 = 3l - 12 \rightarrow l = 22$
$kl = (-3)(22) = -66$

11. (D) 9
If $1/m < 1/n$, then $n < m$. Since m and n are
consecutive integers, then $n + 1 = m$ and
$n = m - 1$. So, $m + (m - 1) = 17 \rightarrow 2m - 1 = 17$
$\rightarrow 2m = 18 \rightarrow m = 9$

12. (A) 42
$(de + m)^2 = 9e^2 + ne + 49$
$d^2e^2 + 2mde + m^2 = 9e^2 + ne + 49$, thus
$d^2 = 9$; $m^2 = 49$; and $ne = 2mde$,
so $d = 3$, $m = 7$, and $n = 2md$
$n = (2)(7)(3) = 42$

13. (D) 5
$14 - 3a < 23 \rightarrow 14 - 23 < 3a \rightarrow -9 < 3a \rightarrow$
$-3 < a$
$a - 5 < -2 \rightarrow a < -2 + 5 \rightarrow a < 3$
So $-3 < a < 3$, so a can be -2, -1, 0, 1, 2 for a
total of **5 integers**.

14. (A) c^{-4}
$b^{1/2} = c^{2/3} \rightarrow b^{(1/2)(4)} = c^{(2/3)(4)} \rightarrow b^2 = c^{8/3}$
$a^{-2/3} = c^{8/3} \rightarrow a^{-2} = c^8 \rightarrow a = c^{-4}$

15. (B) $(6\sqrt{\pi}) / \pi$

If the diagonal of the square equals $6\sqrt{2}$, then the sides of the square measure 6 and the area of the square is 36.

Since the areas of the square and circle are equal, then $\pi r^2 = 36$.

$\pi r^2 = 36 \rightarrow r^2 = 36/\pi \rightarrow r = 6/\sqrt{\pi} \rightarrow$
$r = (6\sqrt{\pi}) / \pi$

16. (E) 10

$$\frac{a^2 - a - 20}{a - 5} = b - 6 + a$$

$(a - 5)(a + 4) / (a - 5) = b - 6 + a \rightarrow$
$a + 4 = b - 6 + a \rightarrow b = 10$

17. (C) 20

$-5x + 8 = 4\sqrt{x} - u \rightarrow (-5)(4) + 8 = 4(\sqrt{4}) - u \rightarrow$
$-20 + 8 = 8 - u \rightarrow -12 = 8 - u \rightarrow -u = -20 \rightarrow$
$u = 20$

18. (A) 1

$(.40)(.60)(x) = (.40)(.10)(y) \rightarrow .24x = .04y \rightarrow$
$6x = 1y$
Product of Ratios $\rightarrow (1:6)(6:1) = 1$

19. (B)

Since the x-value changes by -1, the graph shifts right one unit.
The y-value changes by -2, so the graph shifts down two units.

20. (D) 2.5

$f(x) = (x^2/6) - 4$, so $f(12) = 12^2/6 - 4 \rightarrow$
$f(12) = 20$
$2a(b) = (f(12) + 5)^{1/2} \rightarrow 2a(b) = (20 + 5)^{1/2} \rightarrow$
$2a(b) = 5 \rightarrow a(b) = 2.5$

Section 2

1. (D) $31

Allowance = $15
Homework = (8 ass.)($2/ass.) = $16
So the total money = $15 + $16 = **$31**

2. (E) 1

Only 1 fits: $q^2 = q^3 \rightarrow 1^2 = 1^3 \rightarrow 1 = 1$

3. (E) 100

$(83 + 92 + 79 + 96 + x) / 5 = 90 \rightarrow$
$(350 + x) / 5 = 90 \rightarrow 350 + x = 450 \rightarrow x = 100$

4. (E) 24

m increases by a factor of 4, from 3 to 12.
So n increases by a factor of 4, from 6 to **24**.

5. (D) $y = 3$

$y = mx + b \rightarrow 3 = (m)(-2) + b$; Since the line is perpendicular to the y-axis, the slope (m) is zero.
So $3 = (0)(-2) + b$.
Solve for $b \rightarrow b = 3$, so the equation is $y = 3$.

6. (B) 3

s decreases from 9 to 6, or by a factor of 1.5, so p^2 must increase by the same factor $\rightarrow$
6 to $(6)(1.5) = 9$
$p = p^2 = \sqrt{9} = 3$

7. (A) 1/4

$4k / k^4 = 256 \rightarrow 4 / k^3 = 256 \rightarrow 4 = 256k^3 \rightarrow$
$1/64 = k^3 \rightarrow k = 1/4$

8. (E) 9
Since $h(d) = 3$, the y-value on the graph is 3.
When $y = 3$ on the graph, the x-value equals -3.
$f(x) = 2h(d) - d \rightarrow f(x) = 2(3) - (-3) \rightarrow f(x) = 9$

9. 192
$h(s) = 60s - 3s^2$, so $h(4) = (60)(4) - 3(4)^2 \rightarrow$
$240 - 48 = 192$

10. 0
The total sum of the integers must be 15. Three
people are choosing the integers, so the least
non-negative integer one of them could choose
is zero: 0, 0, 15.

11. $4 / y^2$
$\sqrt{16x} = 8y^{-1} \rightarrow 16x = (8y^{-1})(8y^{-1}) \rightarrow$
$16x = 64/y^2 \rightarrow x = 4/y^2$

12. w
$x = 1$ and $w = -1$, so $x^2/w = 1^2/-1 = -1$.
The point that equals -1 is w.

13. 16
Pick real numbers for b and c: $b = 3$ and $c = 4$
So, $a = b^2c^2 = (3)^2(4)^2 = 144$
b and c doubled $= (6)^2(8)^2 = 2304$
So the value of a is multiplied by 2304/144 =
16

14. 50%
Percent of Change = (New Quantity - Original
Quantity) / Original Quantity × 100%
= (30 – 20) / 20 x 100% = 10/20 x 100% =
(.5)(100%) = **50%**

15. 49
Because the hypotenuses are both $7\sqrt{2}$ and they
form a square, both triangles are 45-45-90. So
their legs, which are also the sides of the square
are each 7. Thus the area of the square is (7)(7)
= **49**

16. II only
The rule $b > b^2 > b$ means that b must be a
positive fraction.

II $\rightarrow (1/2)^3 - (1/2)^2 < (1/2) - ((1/2)^2(1/2)^3) \rightarrow$
$(1/8) - (1/4) < (1/2) - (1/32) \rightarrow -1/8 < 15/32 \rightarrow$
True

17. 228
Winston $\rightarrow$ 4 pies/60min
Clyde $\rightarrow$ 6 pies/60min
1st Hour $\rightarrow$ 4 pies baked in 60 minutes
After 1st Hour $\rightarrow 4/60x + 6/60x = 28 \rightarrow$
$1/6x = 28 \rightarrow x = 168$
Total Time $\rightarrow$ 60 minutes + 168 minutes =
228 minutes

18. 18
The total area of the large square is (6)(6) = 36
The simplest way to begin from there is to "flip
up" the bottom half of the square to combine all
of the shaded regions together. If you do so,
you will realize that the shaded area is exactly
one-half of the large square, or 36/2 = **18**

PSAT MATH

PRACTICE TEST 4

Time-25 Minutes
20 Questions

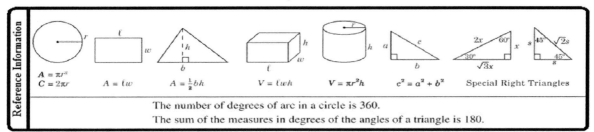

The number of degrees of arc in a circle is 360.
The sum of the measures in degrees of the angles of a triangle is 180.

1. Eunice is the tenth tallest student and the tenth shortest student in the class. How many total students are in the class?

(A) 18
(B) 19
(C) 20
(D) 21
(E) 22

2. In the xy-plane, segment AB passes through points (-4, 4) and (6, y). If the midpoint of the segment is (1, 6), what is the value of y?

(A) 1
(B) 2
(C) 7
(D) 8
(E) 14

3. What is the value of the sum of the internal angles of a triangle subtracted from the sum of the internal angles of a pentagon?

(A) 0
(B) 90
(C) 180
(D) 360
(E) 580

4. A board has six equal spaces, numbered 0-5. If darts are thrown at the board randomly, what is the probability that the darts will <u>not</u> land on an even number on two consecutive throws?

(A) 1/9
(B) 1/4
(C) 4/9
(D) 5/9
(E) 3/4

5. If $\sqrt{x} = x^{-1} = \sqrt{x^2}$, and $x > 0$, then what is the value of x?

(A) 1/2
(B) 1
(C) 2
(D) 4
(E) 9

6. A house has a total of four doors. If Tran cannot enter and exit the house through the same door, how many different ways can she enter and exit the house?

(A) 4
(B) 7
(C) 8
(D) 12
(E) 24

---

7. If the average of a and b is 20, the average of y and z is 20, and the average of k and l is 20, what is the average of a, b, y, z, k, and l?

(A) 20
(B) 24
(C) 25
(D) 30
(E) 40

8.

$$3125, 625, \ldots$$

In the sequence above, each term after the first term, 3125, is one-fifth of the preceding term. Which term is the first term that is <u>not</u> an integer?

(A) 5th term
(B) 6th term
(C) 7th term
(D) 8th term
(E) 9th term

9. If x is proportional to y^2 and $x = -1/2$ when $y^2 = 2$, then what is the value of y when $x = -4$?

(A) 1/4
(B) 1/2
(C) 4
(D) 8
(E) 16

---

10. The smallest integer in a set of consecutive integers is -23. If the total sum of these integers is 75, then how many integers are in the set?

(A) 47
(B) 49
(C) 50
(D) 51
(E) 75

---

11. If 30 percent of 50 percent of q is 60 percent of 75 percent of r, then what is the value of q in terms of r?

(A) $1/3r$
(B) r
(C) $2r$
(D) $3r$
(E) r^2

---

12. If $\dfrac{1}{2}s = t$, $\dfrac{1}{3}t = 6u$, and $9u = 3v$, then what is the value of s in terms of v?

(A) $2v$
(B) $6v$
(C) $12v$
(D) $18v$
(E) $24v$

13.

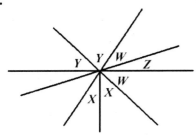

In the figure above, what is the value of *z* in terms of *x*?

(A) $90 - 3x$
(B) $90 - 6x$
(C) $180 - 2x$
(D) $180 - 3x$
(E) $180 - 6x$

14. If $-2x = y^{-1/2}$ and $y^{-1} = (16a^2)$, what is the value of *x* in terms of *a*?

(A) $-8a^2$
(B) $-2a$
(C) $2a^2$
(D) $4a$
(E) $4a^2$

15. If a segment with a slope of two passes through points (-5, 2) and (-3, -*y*), what is y^2?

(A) -6
(B) -4
(C) 6
(D) 18
(E) 36

16.

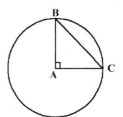

Circle A and triangle ABC are in the figure above. If the length of the hypotenuse of ABC is $4\sqrt{2}$, what is the area of circle A?

(A) 2π
(B) 4π
(C) 8π
(D) 16π
(E) 64π

17. If $h(m) = \dfrac{m^2}{3} - 3$ and $2f(p) = h(6) + 5$, what is the value of *f*(*p*)?

(A) 7
(B) 8
(C) 12
(D) 14
(E) 16

18. How many possible values exist for *k* if $5 - 3k < 14$, $k^{1/2} + 8 < 12$, and *k* is an integer?

(A) 5
(B) 6
(C) 17
(D) 18
(E) 20

19. Max ordered a glass of ice tea with ice cubes. The glass was a perfect cylinder with a height of 8 inches and a diameter of four inches. The glass was filled to the brim with ice tea and 4 ice cubes. If Max decided to take out the ice cubes, and the height of the drink was then 6 inches, what is the volume of one ice cube?

(A) 2π
(B) 4π
(C) 8π
(D) 24π
(E) 32π

20.

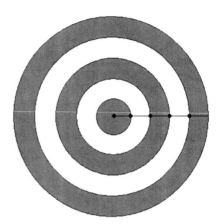

In the figure above, the diameter of the innermost circle is 4. If the diameter of each of the circles is one-half of that of the next largest circle, what is the sum of the circumferences of the five total circles?

(A) 20π
(B) 36π
(C) 60π
(D) 80π
(E) 100π

STOP SECTION

Time-25 Minutes
18 Questions

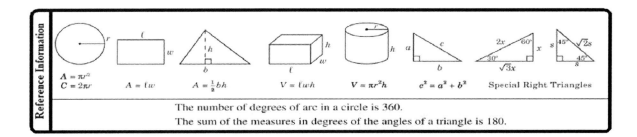

Reference Information

$A = \pi r^2$
$C = 2\pi r$ $A = \ell w$ $A = \frac{1}{2}bh$ $V = \ell w h$ $V = \pi r^2 h$ $c^2 = a^2 + b^2$ Special Right Triangles

The number of degrees of arc in a circle is 360.
The sum of the measures in degrees of the angles of a triangle is 180.

1. If $(2 - k)^2 = 16$, what is the smallest possible value of k^2?

(A) -2
(B) 4
(C) 6
(D) 16
(E) 36

2. If $c^4 = 5$, what is the value of c^{16}?

(A) 20
(B) 25
(C) 60
(D) 125
(E) 625

3. If 60 percent of 20 percent of x percent is 30 percent of 20 percent of y, what is the value of x in terms of y?

(A) .2y
(B) .5y
(C) 20y
(D) 50y
(E) 500y

4. In the xy-plane, P is (3, 4) and Q is (-5, 6). What is the midpoint of segment PQ?

(A) (-1, 4)
(B) (-1, 5)
(C) (4, -1)
(D) (5, -1)
(E) (8, -2)

5.

$$-3, 6, 3,\ldots$$

Each odd term (*ex.* -3) is multiplied by -2 to find the next term. Each even term (*ex.* 6) is subtracted by 3 to find the next term. What is the absolute value of the 12th term?

(A) -126
(B) -63
(C) 63
(D) 126
(E) 129

6. If *a* is inversely proportional to b^{-1} and $a = x$ when $b^{-1} = 4y$, then what is the value of *b*, in terms of *y*, when $a = 4x$?

(A) $1/16y$
(B) $1/8y$
(C) $1/y$
(D) y
(E) $16y$

7. If the height and base of a triangle are halved, what is the value of the triangle's area in terms of base (*b*) and height (*h*)?

(A) bh
(B) $2(bh)$
(C) $bh/2$
(D) $bh/4$
(E) $bh/8$

8. A square lies in the *xy*-plane such that its sides are <u>not</u> parallel to the *x*-axis or *y*-axis. What is the product of the slopes of all four of the square's sides?

(A) -2
(B) -1
(C) 0
(D) 1
(E) 2

9. If $\dfrac{5}{3}w = 20$ and $\dfrac{2}{3}w = 20(k-2)$, then what is the value of k?

10. If $\dfrac{x^2 - 3x - 18}{x+3} = y + 10 + x$, what is y?

11. w and x are positive consecutive even integers and $w < x$. If their sum is 70, what is the value of $\sqrt{x} - w$?

12. Daniel drives 30 mph to the toy store and 45 mph back home, taking the same route both ways. If the total trip takes two hours, how many total miles does he drive on the round-trip from home to the store and back?

13. The ratio of a triangle's sides is 4:3:6. If the length of the hypotenuse is 1/4, what is the value of the length of the short leg of the triangle subtracted from the length of the long leg of the triangle?

14.

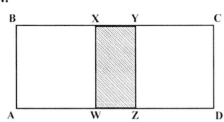

In the figure above, the width of rectangle ABCD is 4 and its length is 12. If the area of rectangle ABYZ is 32 and the area of WXCD is 24, then what is the area of the shaded rectangle WXYZ?

15. If j and k are consecutive odd integers whose sum is 48 and $j > k$, what is the value of $j^{-1/2}$?

16. The function $p(q) = q^2 - 7$ and $f(s) = 9$. If $2f(s) = p(q)$, what is the value of q?

17.

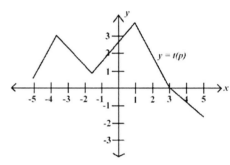

The graph above represents $y = t(p)$. For what range of values of p is $t(p)$ negative?

18.

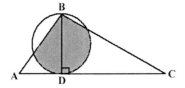

In the figure above, $\triangle ABC$ covers 90% of the circle, as shown by the shaded region that has an area of 87. If the base of $\triangle ABC$ has a length of 13, what is the area of the triangle to the nearest units digit?

STOP TEST

Practice Test 4
Answer Key and Explanations

Section 1

1. B
2. D
3. D
4. C
5. B
6. D
7. A
8. C
9. C
10. C
11. D
12. C
13. E
14. B
15. E
16. D
17. A
18. D
19. A
20. C

Section 2

1. B
2. E
3. D
4. B
5. D
6. C
7. E
8. D
9. 12/5
10. -16
11. -28
12. 72
13. 1/24
14. 8
15. 1/5
16. 5
17. $3 \leq p \leq 5$
18. 128

Finding Your Score

Raw score: Total Number Right – [Total Number Wrong ÷ 4] = _____

Notes: 1) For #9 - #18 in Section 2, only right answers are counted
2) Omissions are not counted towards your raw score
3) If the total number wrong ÷ 4 ends in .5 or .75, it is rounded up

Math Scoring Table			
Raw Score	Scaled Score	Raw Score	Scaled Score
38	80	15	46
37	76	14	45
36	72	13	44
35	70	12	43
34	69	11	42
33	68	10	40
32	66	9	39
31	65	8	38
30	64	7	36
29	63	6	35
28	61	5	34
27	60	4	33
26	59	3	31
25	57	2	29
24	56	1	28
23	55	0	26
22	54	-1	24
21	53	-2	22
20	51	-3	20
19	50	-4	20
18	49	-5	20
17	48	-6	20
16	47	-7	20

Strength and Weakness Review

Go back to the test and circle the questions that you answered incorrectly. This review will allow you to see what answer explanations to study more closely for problem-solving techniques. It will also allow you to see what question types you need to review more carefully.

	Section 1	Section 2
Integers / Consecutive Integers	10	15
Even / Odd Integers		11, 16
Number Lines		
Volume	19	
Sequences	8	5
Exponents	5	2
Fractional / Negative Exponents	14	
Word Problems	1	
Squaring and Square Roots		1
Average, Median, and Mode	7	
Equations with 1 or 2 Variables		9, 10
Equations with 3 or 4 Variables	12	
Inequalities	18	
Building Equations		
Slope	15	8
X-Y Plane Problems	2	4
Functions (Plug-Ins)	17	16
Functions (Graphs & Shifts)		17
Percents	11	3
Ratios		13
Proportions	9	6
Rates		12
Probability	4	
Combinations	6	
Rectangles & Squares		14
Angles	13	
Triangles	3	7
Special Triangles	16	18
Other Polygons		
Circles	16, 20	18

Section 1

1. (B) 19
Since Eunice is the 10th tallest and shortest, there are 9 people shorter and 9 people taller.
$9 + 9 + E = 19$ people.

2. (D) 8
The midpoint is the average of the two points, so $(4 + y) / 2 = 6 \rightarrow 4 + y = 12 \rightarrow y = 8$

3. (D) 360
Internal angles of a pentagon = 540
Internal angles of a triangle = 180
So, $540 - 180 = 360$

4. (C) 4/9
There are four spaces on the board that are not even.
So on the first throw, the probability is 4/6.
On the second throw, the probability is also 4/6.
Two consecutive throws = $(4/6)(4/6) = 16/36 = $ **4/9**

5. (B) 1
Only 1 satisfies all the parts of the equation.
$\sqrt{1}$ (can't be negative) $= 1/1 = \sqrt{1}^2$

6. (D) 12
Since the doors can't be repeated, Tran has four choices of doors by which to enter and she has three choices of doors by which to exit
$(4)(3) = 24$

7. (A) 20
$(a + b + y + z + k + 1) / 6 \rightarrow$
$(40 + 40 + 40) / 6 = 20$

8. (C) 7th term
The pattern is as follows: 3125, 625, 125, 25, 5, 1, 1/5. So the 7th **term** is the first term that is not an integer, or whole number.

9. (C) 4
x increases by a factor of 8 (from -1/2 to -4)
So y^2 increases by a factor of 8, from 2 to 16
If $y^2 = 16$, then $y = 4$

10. (C) 50
The 47 integers from -23 to 23 have a sum of zero. (Make sure to include zero)
The next 3 integers—24, 25, and 26—have a sum of 75.
So there are **50** total integers.

11. (D) $3r$
$(.30)(.50)(q) = (.60)(.75)(r) \rightarrow .15q = .45r \rightarrow$
$q = 3r$

12. (C) $12v$
$9u = 3v \rightarrow u = 3/9v \rightarrow u = 1/3v$
$1/3t = 6u \rightarrow t = 18u \rightarrow t = 18(1/3v) \rightarrow t = 6v$
$1/2s = t \rightarrow s = 2t \rightarrow s = 2(6v) \rightarrow s = 12v$

13. (E) $180 - 6x$
$2y + 3w + 2z + 2x = 360$
Since $w = y$ and $y = 2x$, $10x + 2z + 2x = 360 \rightarrow$
$2z = 360 - 12x \rightarrow z = 180 - 6x$

14. (B) 2*a*

$y^{-1/2} = -2x \rightarrow y^{-1} = (-2x)^2$
$y^{-1} = 16a^2 \rightarrow (-2x)^2 = 16a^2 \rightarrow 4x^2 = 16a^2 \rightarrow$
$x^2 = 4a^2 \rightarrow x = \sqrt{4a^2} \rightarrow$ **$x = 2a$**

15. (E) 36

Slope: $2 = (2 - -y) / (-5 - -3) \rightarrow 2 = (2 + y) / -2$
$\rightarrow -4 = 2 + y \rightarrow y = -6 \rightarrow y^2 =$ **36**

16. (D) 16π

Because they are radii, AB = AC and ABC is
45-45-90
Since BC = $4\sqrt{2}$, AB and AC = 4
Area = $\pi r^2 = \pi(4)^2 =$ **16π**

17. (A) 7

$2f(p) = h(6) + 5 \rightarrow 2f(p) = (6^2 / 3) - 3 + 5 \rightarrow$
$2f(p) = 12 - 3 + 5 \rightarrow 2f(p) = 14 \rightarrow f(p) =$ **7**

18. (D) 18

$5 - 3k < 14 \rightarrow -3k < 9 \rightarrow k > -3$
$k^{1/2} + 8 < 12 \rightarrow \sqrt{k} < 4 \rightarrow k < 16$
$-3 < k < 16$, so k can have **18** possible values.

19. (A) 2π

Volume of Cylinder = $\pi r^2 h = \pi(2^2)8 = 32\pi$
Volume of Ice Tea = $\pi r^2 h = \pi(2^2)6 = 24\pi$
Volume of Ice Cubes = $32\pi - 24\pi = 8\pi$
Volume of one ice cube = $(8\pi) / 4 =$ **2π**

20. (C) 60π

Respective radii = 2, 4, 6, 8, and 10, so
Respective circumferences = 4π, 8π, 12π, 16π,
and 20π
The sum of the circumferences = **60π**

Section 2

1. (B) 4

$(2 - k)^2 = 16 \rightarrow$ The smallest possible value of
k is -2.
So the smallest possible value of $k^2 =$ **4**.

2. (E) 625

$c^4 = 5 \rightarrow c^{(4)(4)} = 5^{(1)(4)} \rightarrow c^{16} = 5^4 \rightarrow c^{16} =$ **625**

3. (D) 50*y*

$(.6)(.2)(x/100) = (.3)(.2)(y) \rightarrow .12x/100 = .06y$
$\rightarrow .12 x = 6y \rightarrow x =$ **50*y***

4. (B) (-1, 5)

Take the average of the two points.
X: $(3 + -5) / 2 \rightarrow -2/2 \rightarrow -1$
Y: $(4 + 6) / 2 \rightarrow 10/2 \rightarrow 5$
(-1, 5)

5. (D) 126

Sequence rules $\rightarrow$
odd terms: -2(odd term)
even terms: even term – 3
Terms: -3, 6, 3, -6, -9, 18, 15, -30, -33, 66, 63,
-126 (12th)
The question asks for $| -126 | =$ **126**

6. (C) 1/*y*

a increases from x to $4x$, or by a factor of 4, so
b^{-1} must decrease by the same factor $\rightarrow$
$(4y)$ to $(4y)(1/4) = y$
$b \rightarrow b^{-1} = y \rightarrow b =$ **1/*y***

7. (E) *bh* **/ 8**

Area = $(1/2)bh = (1/2)(b/2)(h/2) = (1/2)(bh/4) = $ ***bh*/8**

8. (D) 1

The exact value of the slopes cannot be determined from the problem. But two non-parallel sides of the rectangle form a ninety degree angle. This fact means that the slopes are negative reciprocals. So their product must equal -1. Since there are two sets of non-parallel sides, the slope of all four sides of the square is equal to (-1)(-1), or **1**.

9. 12/5

$5/3w = 20 \rightarrow w = (20)(3/5) \rightarrow w = 12$
$2/3w = 20(k-2) \rightarrow 2/3(12) = 20k - 40 \rightarrow$
$8 = 20k - 40 \rightarrow 20k = 48 \rightarrow k = $ **12/5**

10. -16

$$\frac{(x-6)(x+3)}{(x+3)} = y + 10 + x \rightarrow$$

$x - 6 = y + 10 + x \rightarrow y = $ **-16**

11. -28

$w + w + 2 = 70 \rightarrow 2w = 68 \rightarrow w = 34$
$x = 36$ because $w < x$
$\sqrt{x} - w \rightarrow \sqrt{36} - 34 \rightarrow 6 - 34 \rightarrow$ **-28**

12. 72

$x/30 + x/45 = 2 \rightarrow 3x/90 + 2x/90 = 2 \rightarrow$
$5x/90 = 2 \rightarrow 5x = 180 \rightarrow x = 36$
Round trip = $(36)(2) = $ **72**

13. 1/24

$4:3:6 = $ __ : __ : $1/4 \rightarrow 1/6 : 1/8 : 1/4$
$1/6 - 1/8 \rightarrow 4/24 - 3/24 = $ **1/24**

14. 8

Area ABYZ + Area WXCD = 32 + 24 = 56
Area ABCD = (4)(12) = 48
Area WXYZ = 56 – 48 = **8**

15. 1/5

$j - 2 + j = 48 \rightarrow 2j - 2 = 48 \rightarrow 2j = 50 \rightarrow j = 25$
$j^{-1/2} = 25^{-1/2} = 1/\sqrt{25} = $ **1/5**

16. 5

$2f(s) = p(q) \rightarrow (2)(9) = p(q) \rightarrow p(q) = 18$
$p(q) = q^2 - 7 \rightarrow 18 = q^2 - 7 \rightarrow q^2 = 25 \rightarrow q = $ **5**

17. $3 \le p \le 5$

$t(p) = y$, and y is negative on the graph between $p = 3$ and $p = 5$

18. 128

Area of shaded part of circle = (.9)(total area)

So $87 = (.9)(x) \rightarrow 87 = .9x \rightarrow x = 96.67$
Since area of circle = 96.67, $96.67 = \pi r^2 \rightarrow$
$r = \sqrt{96.67} = 9.83$
Thus the diameter = $2r = (2)(9.83) = 19.66$

The diameter is the height of the triangle, and since Area = $bh/2$, the area of triangle ABC = $[(13)(19.66)]/2 = 127.79$, or **128**

PSAT MATH

PRACTICE TEST 5

Time-25 Minutes
20 Questions

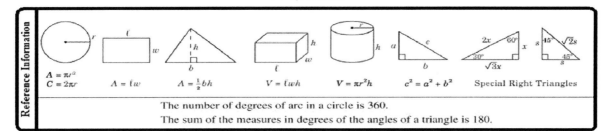

Reference Information

$A = \pi r^2$
$C = 2\pi r$

$A = \ell w$

$A = \frac{1}{2}bh$

$V = \ell w h$

$V = \pi r^2 h$

$c^2 = a^2 + b^2$

Special Right Triangles

The number of degrees of arc in a circle is 360.
The sum of the measures in degrees of the angles of a triangle is 180.

1. If $\dfrac{a+b+c}{3} = \dfrac{a+b}{4}$, then what is the value of c in terms of a and b?

(A) $-a - b$
(B) $7a + 5b$
(C) $(-a - b) / 4$
(D) $(2a + 2b) / 7$
(E) $(7a + 5b) / 4$

2. A mixture contains peanuts, raisins, and cashews in the ratio of 9:7:4. What is the probability of randomly selecting a cashew, expressed as a percentage?

(A) 20%
(B) 25%
(C) 30%
(D) 44%
(E) 50%

3.

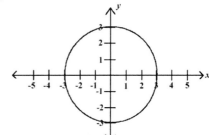

The figure above shows circle O, whose center is the origin. What is the circumference of circle O?

(A) 3π
(B) 6π
(C) 9π
(D) 12π
(E) 36π

4. If $a = b$, $\frac{1}{4}b = 6c$, and $12c = 3d$, then what is the value of a in terms of d?

(A) $\frac{3}{8}d$

(B) $\frac{3}{2}d$

(C) $3d$

(D) $\frac{3}{16}d$

(E) $6d$

5.

In the number line above, which point best approximates $-|e - c|$?

(A) a
(B) b
(C) c
(D) d
(E) f

6. Line l has a slope of 0. If line k is perpendicular to line l, which of the following <u>cannot</u> be true?

(A) Line k has a positive x-intercept
(B) Line k has a negative y-intercept
(C) Line k has a negative x-intercept
(D) Line k has a positive slope
(E) Line k passes through the origin

7. If a segment with a slope of two passes through points $(2, 1)$ and $(z, 9)$, what is the value of z?

(A) -14
(B) -2
(C) 3
(D) 6
(E) 8

8. If $f(x) = -4x + 2x - 2$, then what is the value of $f(2)$?

(A) -6
(B) -4
(C) -2
(D) 2
(E) 4

9. If $(3^3)^2 = 9^x$, then what is the value of 4^x?

(A) 16
(B) 32
(C) 64
(D) 81
(E) 243

10.

$$x, 22, y, z, 646, \ldots$$

In the sequence, each term after the first term, x, is four more than three times the preceding term. What is the value of $x + z$?

(A) 76
(B) 214
(C) 216
(D) 220
(E) 262

11. Gerald drives 30 mph to the park and 50 mph back home, taking the same route both ways. If the total trip takes 4 hours, how many miles does he drive in order to get to the park?

(A) 37.5
(B) 40
(C) 75
(D) 80
(E) 160

12. If $4 = s^t$, $s \leq t$, and s and t are integers, what is the value of $\dfrac{s}{3}$?

(A) 1/3
(B) 2/3
(C) 1
(D) 2
(E) 4

13. If $x = \sqrt{\dfrac{4-y}{\dfrac{1}{8}}}$, what is the value of x when $y^{-1} = 1/2$?

(A) 1/2
(B) 2
(C) 4
(D) 5
(E) 16

14. A board contains four equal areas: red, black, blue, and green. If three darts are thrown at the board, what is the probability that the first dart will <u>not</u> land in the green area, the second dart will land in either the black or the red area, and the third dart will <u>not</u> land in the blue area?

(A) 9/128
(B) 1/64
(C) 9/64
(D) 1/32
(E) 9/32

15.

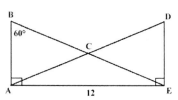

In the figure above, in which AC = CE, what is the length of CD?

(A) $4\sqrt{2}$
(B) $4\sqrt{3}$
(C) $4\sqrt{5}$
(D) $6\sqrt{3}$
(E) $6\sqrt{5}$

16. If s, t, u, v, and w are consecutive integers, where s $< t < u < v < w$, which of the following does <u>not</u> represent the average of the integers?

(A) $\dfrac{s + w}{2}$

(B) $\dfrac{s + t + v + w}{4}$

(C) u

(D) $\dfrac{t + v}{2}$

(E) $\dfrac{s + t}{2}$

17.

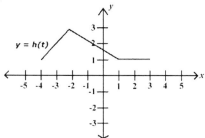

The above graph represents $y = h(t)$. Which of the following graphs best represents the graph of $y = h(t + 1)$?

(A)

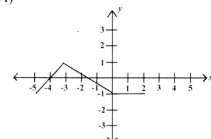

(B)

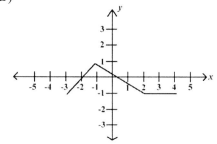

(C)

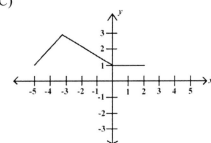

(D)

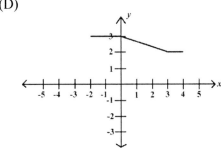

(E)

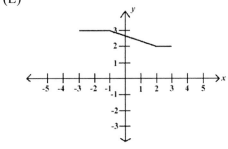

18. A pin code is a sequence of three numbers, each of which is an integer between 0 and 9. If none of the numbers in the sequence can repeat, the first two numbers are greater than five, and the third number must be less than five, then how many combinations for the code are possible?

(A) 12
(B) 24
(C) 30
(D) 60
(E) 80

19.

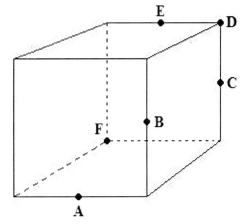

The figure shown above is a cube that has points on its vertexes and edges and point *A* is a midpoint. Which of the following points will create the longest line segment with point *A*?

(A) B
(B) C
(C) D
(D) E
(E) F

20.

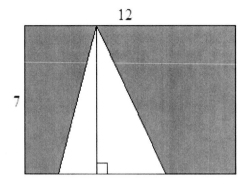

A triangle is inscribed in a rectangle as shown above. If the base of the triangle is x, what is the area of the shaded region in terms of x?

(A) $84 - 7x$

(B) $7x - 84$

(C) $\dfrac{7x}{2} - 38$

(D) $84 - \dfrac{7x}{2}$

(E) $19 - 7x$

STOP SECTION

Time-25 Minutes
18 Questions

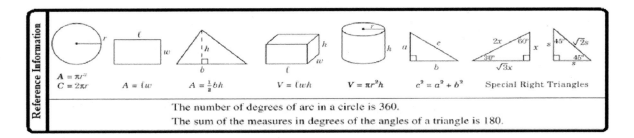

The number of degrees of arc in a circle is 360.
The sum of the measures in degrees of the angles of a triangle is 180.

1. The ratio of a triangle's sides is 3:8:6. If the length of the hypotenuse is 40, what is the value of the length of the short leg subtracted from the length of the long leg?

(A) 3
(B) 10
(C) 15
(D) 25
(E) 30

2. If $-6 < 2 - 3x < 10$, which of the following cannot be a possible value of x ?

(A) -3
(B) -2
(C) -1
(D) 0
(E) 2

3. Line l passes through the points $(1, 3)$ and $(-2, -4)$. What is the reciprocal of the slope of l?

(A) $-\dfrac{7}{3}$
(B) -1
(C) $\dfrac{3}{7}$
(D) 1
(E) $\dfrac{7}{3}$

4. If line w is perpendicular to the x-axis and passes through point $(2, 6)$, what is the equation of line w?

(A) $x = y + 4$
(B) $x = y - 3$
(C) $x = 2$
(D) $x = 6$
(E) $x = -4$

5. If $3 = x^y$, what is the value of $\dfrac{x}{3}$?

(A) x^{1-y}
(B) x^{1+y}
(C) $x^{1/y-1}$
(D) $\dfrac{yx}{3}$
(E) $\dfrac{3}{xy}$

6. Nat drives twelve miles to work each morning, and the same distance back home at night. If she drives 40 mph in the morning and wants her total commute time to be 30 minutes, how fast, in mph, must she drive home from work at night?

(A) 30 mph
(B) 40 mph
(C) 50 mph
(D) 60 mph
(E) 70 mph

7.

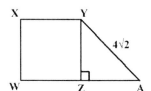

In the figure above, what is the difference of the areas of square WXYZ and triangle ZYA?

(A) 0
(B) 4
(C) 8
(D) 12
(E) 16

8.

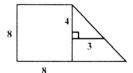

In the figure above, what is the area of the given trapezoid?

(A) 64
(B) 72
(C) 88
(D) 96
(E) 100

9. Original cable rates are increased by 10% and then increased by 30%. What is the percent change from the original to the final price?

10. If $f(x) = 3x + 12$, what is the value of $f(2x)$ in terms of x?

11. Joe is now one year older than his sister was four years ago. If his sister is now 29 years old, how old was Joe four years ago?

12. An equilateral triangle has a perimeter of 21. If the length of a rectangle is equal to the perimeter of that triangle, and the area of the rectangle is 63, what is the width of the rectangle?

13. Ben eats x cookies on Saturday, three times that number on Sunday, and half as many on Monday as on Sunday. What is the average number of cookies that Ben eats each day in terms of x?

14. The first term of a sequence is 100. If each term after the first two terms are two more than one-half the preceding term, what is the fourth term in the sequence?

15.

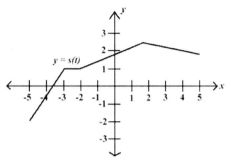

The graph above represents $y = s(t)$. If $s(t) = 1$, what is the range of possible values for t?

16. A circle and square always have the same area. If the radius of the circle is halved, by what factor is the length of the sides of the square reduced?

17. If $a \odot b = 3b - 4a$, what is the value of $(3 \odot 7)^{1/2 \odot 1/2}$?

18. Gregory has a lamp that has 5 different light bulbs. When he turns the knob once, only one light bulb turns on. If he turns the knob again, three light bulbs turn on. The third time he turns the knob, four light bulbs turn on. The fourth time he turns the knob, five light bulbs turn on. The fifth time he turns the knob, none of the light bulbs are on. If he turns the knob 243 times, how many light bulbs are turned on?

TEST END

Practice Test 5
Answer Key and Explanations

Section 1

1. C
2. A
3. B
4. E
5. B
6. D
7. D
8. A
9. C
10. D
11. A
12. B
13. C
14. E
15. B
16. E
17. C
18. D
19. C
20. D

Section 2

1. C
2. A
3. C
4. C
5. A
6. D
7. C
8. C
9. 143%
10. $6x + 12$
11. 22
12. 3
13. $(11x)/6$
14. 16
15. $-3 \leq t \leq -2$
16. 1/2
17. 1/3
18. 4

Finding Your Score

Raw score: Total Number Right – [Total Number Wrong ÷ 4] = _____

Notes: 1) For #9 - #18 in Section 2, only right answers are counted
2) Omissions are not counted towards your raw score
3) If the total number wrong ÷ 4 ends in .5 or .75, it is rounded up

Math Scoring Table			
Raw Score	Scaled Score	Raw Score	Scaled Score
38	80	15	46
37	76	14	45
36	72	13	44
35	70	12	43
34	69	11	42
33	68	10	40
32	66	9	39
31	65	8	38
30	64	7	36
29	63	6	35
28	61	5	34
27	60	4	33
26	59	3	31
25	57	2	29
24	56	1	28
23	55	0	26
22	54	-1	24
21	53	-2	22
20	51	-3	20
19	50	-4	20
18	49	-5	20
17	48	-6	20
16	47	-7	20

Strength and Weakness Review

Go back to the test and circle the questions that you answered incorrectly. This review will allow you to see what answer explanations to study more closely for problem-solving techniques. It will also allow you to see what question types you need to review more carefully.

	Section 1	Section 2
Integers / Consecutive Integers	16	
Even / Odd Integers		
Number Lines	5	
Absolute Values		
Sequences	10	14
Exponents	9	5
Fractional / Negative Exponents	13	
Word Problems		11, 18
Squaring and Square Roots	13	
Average, Median, and Mode	16	
Equations with 1 or 2 Variables	12	
Equations with 3 or 4 Variables	1, 4	
Inequalities	6	2
Building Equations		13
Slope	7	3
X-Y Plane Problems		4
Functions (Plug-Ins)	8	10
Functions (Graphs & Shifts)	17	15
Percents		9
Ratios		1
Proportions		
Rates	11	6
Probability	2, 14	
Combinations	18	
Rectangles & Squares	20	7, 12, 16
Angles		
Triangles	15, 20	12
Special Triangles	15	7
Other Polygons		8, 16
Circles	3	
Special Symbols		17
Solid Geometry	19	

Section 1

1. (C) $(-a - b) / 4$
Cross-Multiply $\rightarrow 4a + 4b + 4c = 3a + 3b \rightarrow$
$4c = -a - b \rightarrow c = (- a - b) / 4$

2. (A) 20%
Based on the proportion (9:7:4), cashews represent 4 parts out of the 20 total parts. As a fraction this equals 4/20 or **20%.**

3. (B) 6π
Since the radius of the circle is three, the circumference is $2\pi(3) = 6\pi$

4. (E) $6d$
$12c = 3d \rightarrow 6c = (3/2)d$
$(1/4)b = 6c \rightarrow (1/4)b = (3/2)(d) \rightarrow b = 6d$
$a = b = 6d$

5. (B) point b
point $e = \sim 0.5$ and point $c = \sim -0.1$
$-| e - c | \rightarrow -| 0.5 - (-0.1) | \rightarrow -| 0.5 + 0.1 | \rightarrow$
$-| 0.6 | \rightarrow 0.6$

6. (D) Line k has a positive slope

Line l has a slope of 0. Line k is perpendicular to line l and perpendicular lines have slopes that are negative reciprocals. The negative reciprocal of 0 is 0, so line k <u>cannot</u> have a positive slope.

7. (D) 6
$2 = (9 - 1) / (z - 2) \rightarrow 2(z - 2) = 8 \rightarrow 2z - 4 = 8$
$\rightarrow z = 6$

8. (A) -6
$f(x) = -4x + 2x - 2 \rightarrow f(x) = -2x - 2$
$f(2) = -2(2) - 2 \rightarrow f(2) = -4 - 2 \rightarrow f(2) = -6$

9. (C) 64
$(3^3)^2 = 9^x \rightarrow 3^6 = 3^{2x} \rightarrow 2x = 6 \rightarrow x = 3$
$4^x = 4^3 = 64$

10. (D) 220
$x \rightarrow 22 = 3x + 4 \rightarrow 3x = 18 \rightarrow x = 6$
$z \rightarrow 646 = 3z + 4 \rightarrow 3z = 642 \rightarrow z = 214$
$x + z = 220$

11. (A) 37.5
$x/30 + x/50 = 4 \rightarrow 5x/150 + 3x/150 = 4 \rightarrow$
$8x/150 = 4 \rightarrow 8x = 600 \rightarrow x = 75$

Since the question asks the distance only to the park, the total distance, x, is divided by two$\rightarrow$ **37.5**

12. (B) 2/3
$4 = s^t \rightarrow$ The only numbers that work under the given conditions are $s = 2$ and $t = 2$, so $s/3 = 2/3$

13. (C) 4
$y^{-1} = \frac{1}{2} \rightarrow y = 1/(1/2) \rightarrow y = 2$
$x = \sqrt{(4 - 2)/(1/8)} \rightarrow \sqrt{2/(1/8)} \rightarrow \sqrt{16} \rightarrow 4$

14. (E) 9/32
1st Event $\rightarrow$ 3/4
2nd Event $\rightarrow$ 1/2
3rd Event $\rightarrow$ 3/4
Probability of Consecutive Events =
$(3/4)(1/2)(3/4) = 9/32$

15. (B) 4√3

Step 1

<AEC = 180 – 90 – 60 = 30, so <DEC = 90 – 30 = 60

AC = CE, so <CAE is 30, and <EDA is 180 – 90 – 30 = 60

Since <DEC = 60 and < EDA = 60, <DCE = 60

Thus ΔADE is 30-60-90 and ΔCDE is equilateral

Step 2

In Δ ADE, length 12 corresponds to length $x\sqrt{3}$ in the 30-60-90 triangle, so $x\sqrt{3} = 12 \rightarrow x = 12/\sqrt{3} \rightarrow x = (12\sqrt{3})/3 \rightarrow x = 4\sqrt{3}$

In ΔADE, DE corresponds to length x in the 30-60-90 triangle, so DE = x = 4√3

Step 3

Since Δ CDE is equilateral, DE = CD = **4√3**

16. (E) (s + t) / 2

Five consecutive integers: $s, t, u, v, w \rightarrow$

plug-in: 2, 3, 4, 5, 6

Average of integers = median = 4

$(s + t) / 2 = (2 + 3) / 2 = 5/2 \rightarrow 5/2 \neq 4$, so

(s + t) / 2

17. (C)

Since the *x*-value changes by 1, the graph shifts left one unit.

18. (D) 60

categories = 3

options → 1st # = 4 choices (# > 5)

→ 2nd # = 3 choices (# can't repeat, # > 5)

→ 3rd # = 5 choices (#s can't repeat, # < 5)

Total # of combinations = (4)(3)(5) = **60**

19. (C) *D*

AD will be the longest line segment because point D is the furthest away in all 3 dimensions: height, length, and width.

20. (D) $84 - \dfrac{7x}{2}$

Area of Rectangle = lw = (7)(12) = 84

Height of triangle = 7 and base = x, so

Area of triangle = $(7x)/2$

So area of shaded region = **84 – (7x/2)**

Section 2

1. (C) 15

3 : 8 : 6 → __ : 40 : __ → 15 : 40 : 30

30 – 15 = **15**

2. (A) -3

$-6 < 2 – 3x < 8 \rightarrow -6 < 2 – (3)(-3) < 8 \rightarrow$

-6 < 11 < 8

Since this is not true, **-3** cannot be the value of x.

3. (C) 3/7

$(y_2 – y_1) / (x_2 – x_1) \rightarrow (-4 – 3) / (-2 – 1) \rightarrow$ -7/-3

→ slope = 7/3

reciprocal of slope = **3/7**

4. (C) x = 2

Because line w is perpendicular to the x-axis where $x = 2$, it will always be a line for which $x = 2$.

5. (A) x^{1-y}

$3 = x^y$, so $x/3 = x^1/x^y \rightarrow x/3 = x^{1-y}$

6. (D) 60

Morning $\rightarrow$ 12 miles / 40 mph
Evening $\rightarrow$ 12 miles / x
$12/40 + 12/x = 1/2$ (hour) $\rightarrow 12/x = 1/2 - 12/40$
$\rightarrow 12/x = (20 - 12) / 40 \rightarrow 12/x = 1/5 \rightarrow$
$x = $ **60 mph**

7. (C) 8

Triangle ZYA is a 45-45-90 triangle, so YZ and ZA = 4
Area Triangle= $(1/2)bh = (1/2)(4)(4) = 8$
YZ is also a side length of square WXYZ, so
Area Square = $s^2 = (4)^2 = 16$
Area WXYZ – Area ZYA = $16 - 8 =$ **8**

8. (C) 88

The two triangles are similar, so if the height of the smaller triangle is 4, the height of the larger triangle is 8. Similarly, if the base of the smaller triangle is 3, the base of the larger triangle is 6.
Area Triangle = $(1/2)bh = (1/2)(6)(8)= 24$
Area Square = $s^2 = 8^2 = 64$
Area of Trapezoid = $64 + 24 =$ **88**

9. 143%

x = original price $\rightarrow (1.1x)(1.3) \rightarrow 1.43x \rightarrow$
143% change

10. 6x + 12

$f(x) = 3x + 12 \rightarrow f(2x) = 3(2x) + 12 \rightarrow$
$f(2x) = $ **6x + 12**

11. 22

Now: S = 29, so 4 years ago S = 25
Now: J – 1 = 25, so J = 26, so
Four years ago = J – 4 = 26 – 4 = **22**

12. 3

Rectangle, Area = $wl \rightarrow 63 = (w)(21) \rightarrow w = 3$

13. (11x) / 6

Saturday = x ; Sunday = $3x$; Monday = $3x/2$

$$\frac{x + 3x + \dfrac{3x}{2}}{3} = \frac{x}{3} + x + \frac{x}{2} = \frac{2x + 6x + 3x}{6} \rightarrow$$

(11x)/ 6

14. 16

1st Term: 100; 2nd Term: 52; 3rd Term: 28;
4th Term: 16

15. -3 $\leq t \leq$ -2

Since $s(t) = 1$, the y-value on the graph is 1. When $y = 1$ on the graph, the x-value can only have a value between -2 and -3, so **-3 $\leq t \leq$ -2**

16. 1/2
Area of Old Circle (πr^2) = Area of Old Square
Area of New Circle (y) = $\pi(r/2)^2$ = $(\pi)(r^2/4)$
= $\pi r^2 / 4$, so
Area New Square (y) → $s^2 / 4$
Side Length New Square (s) = $s^2 / 4$ →
$(\sqrt{s})^2 / (\sqrt{4})$ → $s / 2$
So, the side lengths of the square are reduced by
1/2.

17. 1/3
For the base, $a = 3$ and $b = 7$, so $3b - 4a$ →
$(3)(7) - (4)(3)$ → $21 - 12$ → 9
For the exponent, $a = 1/2$ and $b = 1/2$, so
$(3)(1/2) - (4)(1/2)$ → $3/2 - 2$ → $-1/2$
Together: $9^{-1/2} = 1/\sqrt{9}$ → **1/3**

18. 4
There are ($243/5$) or 48 full 5-turn sequences,
which can be disregarded. Since there is a
remainder of three, however, Gregory stops
turning the knob after the 3rd time.
So when he stops, there are four lights on.

AVAILABLE TITLES FROM FUSION PRESS

5 SAT Math Practice Tests

5 SAT Critical Reading Practice Tests

5 SAT Writing Practice Tests

10 SAT Vocabulary Practice Tests

Score-Raising Vocabulary Builder

FORTHCOMING TITLES FROM FUSION PRESS

5 PSAT Critical Reading Practice Tests

5 PSAT Writing Practice Tests

5 SAT Math Practice Tests for the Advanced Student

5 SAT Critical Reading Practice Tests for the Advanced Student

5 SAT Writing Practice Tests for the Advanced Student

5 ACT Math Practice Tests

CPSIA information can be obtained at www.ICGtesting.com
Printed in the USA
BVOW050322060313

314838BV00001B/9/P

9 780979 678660